Argishti Melikyan

Active and Passive Plasmonic Devices for Optical Communications

Karlsruhe Series in Photonics & Communications, Vol. 17
Edited by Profs. J. Leuthold, W. Freude and C. Koos

Karlsruhe Institute of Technology (KIT)
Institute of Photonics and Quantum Electronics (IPQ)
Germany

Active and Passive Plasmonic Devices for Optical Communications

by
Argishti Melikyan

KIT Scientific Publishing

Karlsruher Institut für Technologie
Institut für Photonik und Quantenelektronik

Active and Passive Plasmonic Devices for Optical Communications

Zur Erlangung des akademischen Grades eines Doktor-Ingenieurs von der KIT-Fakultät für Elektrotechnik und Informationstechnik des Karlsruher Instituts für Technologie (KIT) genehmigte Dissertation

von Argishti Melikyan, M. Sc. geboren in Aschtarak, Armenien

Tag der mündlichen Prüfung: 9. Mai 2014
Referent: Prof. Dr. sc. nat. Jürg Leuthold
Korreferent: Prof. Dr.-Ing. Dr. h. c. Wolfgang Freude

Impressum

KIT Scientific Publishing

Karlsruher Institut für Technologie (KIT)
KIT Scientific Publishing
Straße am Forum 2
D-76131 Karlsruhe

KIT Scientific Publishing is a registered trademark
of Karlsruhe Institute of Technology.
Reprint using the book cover is not allowed.

www.ksp.kit.edu

Print on Demand 2018 – Gedruckt auf FSC-zertifiziertem Papier

ISSN 1865-1100
ISBN 978-3-7315-0463-4
DOI 10.5445/KSP/1000051316

Table of Contents

Kurzfassung

Plasmonik ist eine Technik, um Licht mit Hilfe metallischer Leiter zu beeinflussen. Plasmonische Bauteile gehören zu einer neuen Generation höchst-integrierter photonischer Schaltungen, deren Abmessungen weitaus kleiner als die Lichtwellenlänge sein können. Die verfügbare Chipfläche wird dadurch optimal genutzt, was die Produktionskosten zukünftiger optoelektronisch integrierte Schaltkreise (integrated circuit, IC) dramatisch reduzieren könnte. Darüber hinaus reduziert die Plasmonik den Energieverbrauch optoelektronischer Grundelemente der Telekommunikation wie Modulatoren, Schalter und Router. Zusätzlich bieten plasmonische Modulatoren und Schalter aufgrund ihrer kompakten Abmessungen und ihrer ultra-kleinen RC-Zeitkonstanten höchste Schaltgeschwindigkeiten.

Diese Arbeit beschäftigt sich mit den Grundlagen plasmonischer optischer Schaltungen und konzentriert sich auf die Entwicklung von neuartigen, plasmonischen Modulatoren. In den entwickelten Bauteilen wird die optische Phase oder die Ausbreitungsdämpfung von Oberflächen-Plasmon-Polaritonen (surface plasmonic polariton, SPP) durch eine angelegte elektrische Spannung moduliert. Wir zeigen zwei verschiedene Konzepte plasmonischer Modulatoren, eines, um die Phase der SPP zu beinflussen, und eines zur Veränderung deren Intensität.

Um die SPP-Phase zu modulieren, verwenden wir einen plasmonischen Wellenleiter, der mit einem nichtlinearen Dielektrikum gefüllt ist und das über den Pockels-Effekt-auch bekannt als linearer elektrooptischer Effekt-die Ausbreitunggeschwindigkeit eines SPP verändert. Zum ersten Mal konnten wir zeigen, dass ein plasmonischer Phasenmodulator, integriert auf einem Silizium-Chip, einem optischen Träger Datenraten von 40 Gbit/s aufmodulieren kann.

Anschließend erörtern wir das Konzept eines SPP-Absorptionsmodulators mit einer Länge von nur wenigen Mikrometern. Wir zeigen, dass mit dünnsten Schichten von transparenten Metalloxiden wie Indium-Zinn-Oxid (indium tin oxide, ITO) ultra-kompakte und damit hochgeschwindigkeitstaugliche Intensitätsmodulatoren realisiert werden können.

Um die optischen Einfügeverluste der plasmonischen Strukturen zu minimieren, entwickeln wir effiziente Übergänge (Modenkonverter), welche die effiziente Kopplung eines SPP an den Grundmodus eines Streifenwellenleiters aus Silizium gestattet. Darüber hinaus entwickeln wir einen speziellen Modenwandler, der polarisationsselektiv sowohl den TE- als auch den TM-Modus eines Silizium-Streifenwellenleiters in das entsprechende SPP konvertiert. Bei Kombination eines plasmonischen Modulators mit einem der oben erwähnten Modenwandler kann einem optischen Träger ein Datenstrom in einer der beiden Polarisationen aufmoduliert werden. Werden zwei orthogonale Polaristation in einem Polarisations-Multiplex zur Übertragung genutzt, kann man die Datenrate einer optischen Verbindung zwischen zwei silizium-photonischen Schaltungen verdoppeln.

Polarisationsmultiplex erfordert die Zusammenführung und die Trennung orthogonaler Poarisationen, was mit einem Polarisationsstrahlteiler (polarization beam splitter, PBS) erreicht werden kann. Wir entwickeln einen ultra-kompakten plasmonischen PBS, dessen Länge kleiner als 5 µm ist und der die TE- und TM-Moden eines Streifenwellenleiters aus Silizium trennt oder vereint. Plasmonische PBS werden sowohl im Sender als auch im Empfänger genutzt.

In zwei Anhängen werden einerseits die mathematischen Grundlagen der Arbeit erläutert und andererseits die Herstellung der Bauteile sowie die Experimente im Detail geschildert.

Preface

Plasmonics is a technology that uses metals to manipulate the electromagnetic waves at optical frequencies. Plasmonics is considered to be an alternative path towards a new generation of more densely integrated photonic circuits with dimensions that would not be possible with any other conventional photonic technology [1, 2]. The small footprint of plasmonic devices can reduce the production costs of current optoelectronic integrated circuits (ICs). In addition, plasmonics can have a great impact on minimizing the energy consumption of optoelectronic devices such as modulators, switches and routers, which play a major role in the field of telecommunications. Furthermore, plasmonic modulators and switches inherently offer high-speed operation because of their compact dimensions and of their ultra-small RC time constants due to the high conductivity of metals.

In this thesis, we focus on the development of novel plasmonic modulators, wherein the surface plasmon polariton (SPP) is modulated by an electrical signal. We demonstrate two concepts of plasmonic modulators for controlling the phase and the intensity of the SPP. First, we use the Pockels effect (also known as linear electro-optic effect) in electro-optic organic materials to modulate the phase of the SPP. We demonstrate 40 Gbit/s operation of plasmonic phase modulators (PPM) and Mach-Zehnder interferometric (MZI) modulator integrated on a silicon-on-insulator photonic chip. Then, we introduce the idea of an SPP absorption modulator with a length of only few micrometres. We show that ultra-compact and high speed SPP intensity modulators can be realized by using thin layers of transparent metal oxide such as indium tin oxide (ITO).

To minimize optical insertion losses in plasmonic modulators we develop efficient schemes to excite the modulator's SPP with the fundamental mode of a silicon strip waveguide. Specifically, we design photonic-to-plasmonic mode converters that can be used to convert both TE and TM modes of a silicon strip waveguide into the proper SPP. By combining a plasmonic modulator with the mode converters mentioned above, we can encode a data stream in either of both polarizations of a silicon strip waveguide. The simultaneous transmission of two polarizations in one waveguide doubles the aggregate data rate. To make such a polarization diversity scheme work with silicon photonic circuits, we propose an ultra-compact plasmonic polarization beam splitter (PBS) with a sub-5 µm length. These plasmonic PBSs (de-)multiplex orthogonal polarizations (TE and TM modes) of a silicon strip waveguide.

The thesis is organized as follows. In Chapter 1, the motivation for this work is given, and the state of the art of optical modulators is presented and compared to the plasmonic modulators developed by us.

In the first part of Chapter 2, we give an overview of the basic theory of linear and nonlinear electromagnetic field theory. We discuss the optical properties of metals and give the classical models for the linear susceptibility. In the second part of Chapter 2, we give a brief introduction to the theory of surface plasmon polaritons. The SPP dispersion relations for various metallic structures are discussed, and the most important properties of SPP are explained.

In Chapter 3, we present the passive plasmonic components developed within this thesis. First, we report on a novel photonic-to-plasmonic mode converter, which can efficiently convert a quasi-TM mode of a silicon strip waveguide into the gap SPP of horizontal metallic slot waveguide (MSW). The concept of a metallic taper as a mode converter (MTMC) is introduced in the second part of Chapter 3. We present the optimum configuration of the MTMC mode converter and discuss the performance of the fabricated device. Finally, the operation principle of a novel ultra-compact polarization beam splitter is described and design parameters are reported. The content of Chapter 3 has been published in one journal [J1] and one conference paper [C1].

In Chapter 4, we demonstrate two concepts of plasmonic modulators. First we introduce a plasmonic phase modulator (PPM) and experimentally compare the performance of two generations of fabricated PPMs. Detailed optical and RF characterization results of the second generation (2nG) PPM are presented, and operation in high temperature environments are experimentally confirmed. The operations of the 1stG and 2ndG PPMs at data rates of up to 40 Gbit/s are demonstrated. Finally, we describe the operation principle of a surface plasmon polariton absorption modulator (SPPAM). We show how the plasma dispersion effect in an ultra-thin ITO layer can be used to manipulate the wavelength dependent peak of the SPP absorption spectrum. We show that a device with a length of only a few micrometers is required to attain desired modulation properties, with an estimated RF bandwidth exceeding 100 GHz. Theoretical predictions are validated by a proof-of-principle experiment performed with the fabricated SPPAM. The content of Chapter 4 has been published in two journal contributions [J2, J3] and two conference papers [C2, C3].

In Appendix A, we give the theoretical derivations of the concepts that are essential for understanding, describing and predicting the performance and operation of the plasmonic components designed in this work.

Finally, in Appendix B, the fabrication and experimental details of the investigated plasmonic phase modulators are described.

Achievements of the Present Work

In this thesis, novel plasmonic modulators are designed and demonstrated. The main achievements within the thesis are summarized as:

Surface plasmon polariton phase modulator: A novel high speed plasmonic phase modulator is designed and demonstrated based on the Pockels (electro-optic) effect in an electro-optic organic material. The demonstrated plasmonic phase modulator is integrated with other silicon photonics circuits on the same silicon-on-insulator chip. The modulator occupies an area as compact as $1 \times 29\ \mu m^2$, has ultra-wide RF and optical bandwidths, and can operate at high temperatures. For the first time ever, a plasmonic modulator is demonstrated to work at a bitrate of 40 Gbit/s. An error-free (BER $< 1{\times}10^{-10}$) binary phase shift keying (BPSK) modulation using plasmonic phase modulator is demonstrated at data rates up to 40 Gbit/.

Plasmonic organic hybrid Mach-Zehnder modulator: A novel high-speed plasmonic-organic hybrid (POH) Mach-Zehnder modulator comprising ultra-compact plasmonic phase shifters with lengths as small as 19 µm is demonstrated. On-off keying (OOK) modulation at the data rates of 40 Gbit/s using a POH Mach-Zehnder modualtor is demonstrated. Furthermore, the influence of the length of the plasmonic phase shifter on the quality of the OOK signal is also discussed.

Surface plasmon polariton absorption modulator: A novel plasmonic absorption modulator has been designed and demonstrated on a silicon-on-insulator platform. The modulator can be as short as (1 … 2) µm with an extinction ratio of around 1 dB. The RF bandwidth of the modulator is estimated to be in the THz range, and is limited only by the charging time of the small intrinsic metal-insulator-metal capacitor. Such modulators with a footprint of only a few square micrometers can be used in future high-density integrated optoelectronics.

Photonic-to-plasmonic mode converter: A novel approach has been proposed for converting the silicon strip photonic mode to the gap surface plasmon polariton (SPP) of a horizontal metallic slot waveguide. The scheme provides more than 85 % conversion efficiency for the horizontal metallic slot waveguide having a slot size of (30 … 50) nm. The converter can be used in plasmonic phase modulators with a horizontal metallic slot waveguide configuration.

Plasmonic polarization beam splitter: A novel polarization beam splitter has been proposed for application in silicon photonic circuits exploiting the properties of a horizontal metallic slot waveguide. The device has a length of 3.5 µm and provides an extinction ratio of 10 dB. Such compact polarization beam splitters can be applied in short-haul and long-haul communication systems that employ polarization diversity schemes.

1 Introduction

The global semiconductor industry has been outperforming expectations for the last 40 years by continuously increasing the transistor densities on integrated circuits (IC) [3]. Reducing the single transistor size has resulted in a significant increase in the CPU clock frequency [4]. However, as the feature sizes reach 22 nm, leakage currents become increasingly problematic and limit the speed of a CPU [5, 6]. In this scenario the memory and I/O access times limit the processing rate rather than the outright CPU clock frequency. [7]. In this case the speed limiting factor is not the CPU clock frequency any more, but rather the memory and I/O access times [8]. Current chip-to-chip copper interconnects suffer from bandwidth limitations and from a large energy consumption. High-speed interconnects with the lowest energy consumption can be realized by optical means, which so far have shown to be capable of transmitting signals with a data rate of 26 Tbit /s [9]. However, chip-to-chip optical interconnects require high-speed optical transceivers which need to be monolithically integrated on electronic ICs [10]. Optical transmitters are necessary to convert digital information from the electrical to the optical domain. A typical high-speed optical transmitter comprises an electronic driver, a light source, and an optical modulator that is responsible for electrical-to-optical signal conversion [11]. To reduce the production cost of optical transmitters, the optical modulators should be fabricated with a process compatible with complementary metal oxide semiconductor (CMOS) technology [10]. Silicon photonics is forecast to be the most promising platform for the realization of silicon optical transmitters that can be, monolithically integrated with electronic ICs on the same chip. Silicon modulators should feature small area footprint and the lowest energy consumption to lower the fabrication and the operating costs, respectively [11]. Furthermore, modulators should offer a large RF bandwidth and an operation across a large optical wavelength range around 1550 nm.

In an optical modulator, light is controlled by an electrical signal [11]. This can be done in a waveguide modulator by changing its refractive index. A control of the real part of the refractive index (phase modulator, PM) can be done by the thermo-optic effect [12–17], the electro-optic (EO) effect [18–26], and the plasma dispersion (PD) effect [27–45]. The imaginary part of the refractive index (intensity modulator, IM) is influenced by the electro-absorption (EA) effect [46–53], the plasma dispersion effect [54–57], or by a phase transition of a medium [58–61]. The change of the real part of the refractive index in a liquid crystal [62–64] is useful for phase tuning, but far too slow for a fast data modulation. This is also true for thermo-optic [12–17] and for some phase transition modulators [58]. Only the modulators employing EO, EA and PD can provide modulation speeds up to 40 Gbit/s.

1.1 Intensity modulators (IM)

In IMs, the intensity of light is directly modulated by an electrical signal $U(t)$, see Figure 1.1. The most important characteristics of an IM are the radio frequency (RF) bandwidth for the modulation section, and the modulator's extinction ratio which is defined by the ratio of the high level output optical power P_1 and the low level output power P_0,

$$\mathrm{ER} = 10\lg\left(P_1/P_0\right). \tag{1.1}$$

The RF bandwidth and the extinction ratio of IMs should be maximized to ensure a high-quality optical signal in the presence of noise.

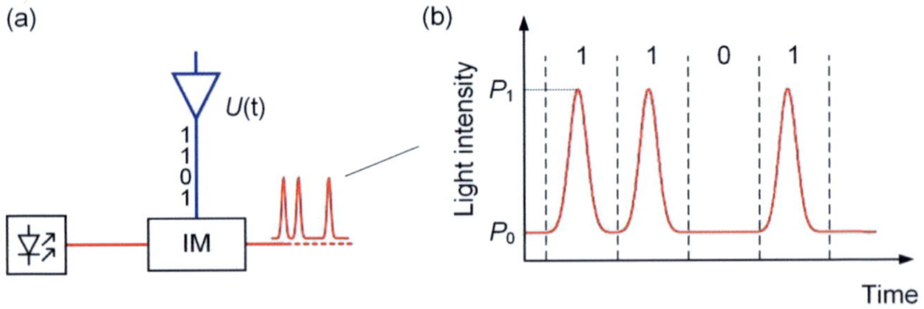

Figure 1.1 Optical transmitter based on the electro-absorption modulator. (a) The electrical and optical parts of the transmitter are depicted in blue and red, respectively. Light intensity is modulated by the electrical signal $U(t)$. (b) The optical signal at the output of the transmitter is characterized by its high and low power levels of P_1 and P_0, respectively.

Compact IMs are realized using the Franz-Keldysh (FK) effect in bulk semiconductors, and the quantum-confined Stark (QCS) effect in quantum wells [65, 66]. These effects (also known as electro-absorption effects) describe the change in an optical absorption of a semiconductor due to the applied electric field [67]. IMs employing either of the electro-absorption effects are named as electro-absorption modulators (EAMs). EAMs with lengths of around 50 μm are realized on silicon by epitaxially growing Ge or GeSi compounds on silicon-on-insulator wafers [46–53]. A device with an ER of 6 dB is reported for a drive voltage as low as $U = 2.8\ V_{pp}$. [49] However, the EAMs show optimum operation only within a limited wavelength range, which, however, has been shown be tunable by varying the amount of Si the in GeSi layer.

IMs on silicon can also be constructed using ring resonator devices, these type of modulators are also known as ring resonator modulators (RRMs) [19, 27, 30, 31, 40, 41, 44]. In RRMs, the resonant wavelength is tuned by modulating the refractive index of the ring. This is typically done by using the plasma dispersion effect in silicon [27–31, 40–43]. The RRMs offer ultra-compact dimensions. For instance, a device with a footprint of $9 \times 9\ \mu m^2$ has been demonstrated based on the plasma dispersion effect through carrier injection [27]. Devices based on the carrier injection process exhibit an RF bandwidth of only few GHz [27, 29], because of the long carrier recombination lifetime. Nevertheless, by pre-emphasizing the electrical drive signal, an RRM can produce a data stream at 50 Gbit/s as has been demonstrated recently [40]. The pre-emphasis is not required in RRMs which employ the plasma dispersion effect due to the fast carrier depletion process in silicon [30, 31, 41–43]. A device with a footprint of $12 \times 12\ \mu m^2$ and an operating speed of 44 Gbit/s is demonstrated using the carrier depletion effect in silicon [42]. RRMs require small drive voltages and have an energy consumption as low as 32 fJ/bit, however, for an extinction ratio of only 1.1 dB [41]. In addition, ring resonators show an optimum operation only near their design wavelengths, therefore, they are quite sensitive with respect to temperature fluctuations, and have a limited RF bandwidth because of the finite photon lifetime in a resonator.

In this thesis, we first report on a compact IM which is realized using plasmonics and surface plasmon polaritons (SPPs) [J4, C4]. Our approach to realizing an

IM is based on the plasma dispersion effect in a metal-oxide layer [54, 68]. We show that ultra-fast IMs with lengths of only 1 … 2 µm and extinction ratios of still 1 dB can be realized by employing a thin layer of indium tin oxide (ITO) in a metal-insulator-metal plasmonic waveguide structure. Using the same effect, other researchers have reported extinction ratios larger than 1 dB for a plasmonic IM comprising a layer stack of silicon-silica-ITO-gold [68]. The high extinction ratio is achieved by modulating the absorption coefficient of a photonic-plasmonic hybrid mode [69] through the plasma dispersion effect in ITO. However, the large resistivity of the intrinsic silicon layer makes the device reported in Ref. [69] useful only for low-speed applications; fast modulation of hybrid SPPs yet to be shown.

1.2 Phase modulators (PM)

For the vast majority of silicon optical modulators, the phase of the light is modulated by an electrical signal [67]. PMs have shown to be tolerant to temperature fluctuations, and to provide RF and optical bandwidths that are larger than the ones of conventional EAMs and RRMs [11].

The strength of the phase modulation $\vartheta(t)$ in a PM is proportional to its length and to the strength of an electrical drive signal $U(t)$ [67]. Hence, there is a compromise between the electrical energy consumption and the device footprint. To accurately evaluate and compare the performance of reported phase modulators, the performance metric of the voltage-length product $U_\pi L$ that results in a static phase shift of $\vartheta = \pi$ is used. A PM that is characterized by a smaller $U_\pi L$ product has a smaller footprint and consumes less energy.

Phase modulators are also building blocks of amplitude or intensity modulators. When PMs are positioned in the arms of a Mach-Zehnder interferometer (MZI), see Figure 1.2, the phase modulation is converted to an intensity modulation at the MZI output. Such a device is known as a Mach-Zehnder modulator (MZM). If the phase modulators are driven with electrical signals $U_1(t)$ and $U_2(t)$, the respective optical phase shifts $\vartheta_1(t)$ and $\vartheta_2(t)$ result. Depending on the phase difference $\vartheta_1(t) - \vartheta_2(t)$, the output intensity of a MZM can vary from maximum transmission (constructive interference) to minimum transmission (destructive interference) as schematically shown in Figure 1.2(b) [70]. The transfer characteristic of the IM has a raised-cosine shape. When the electric field is considered, the transfer characteristic is a sine as in Figure 1.2(c) [70]. The operating point (a relative phase difference between two arms in the case of $\vartheta_1(t) = 0$ and $\vartheta_2(t) = 0$) is defined by adding an extra phase φ to one of the arms, see Figure 1.2(a).

A signal's in-phase (I) and quadrature-phase (Q) component can be modulated with two MZMs nested in the two arms of another MZI (IQ modulator) having a relative phase difference of $\pi / 2$ [71]. IQ modulators are needed for advanced modulation formats, where both amplitude and phase of an optical carrier is varied [71].

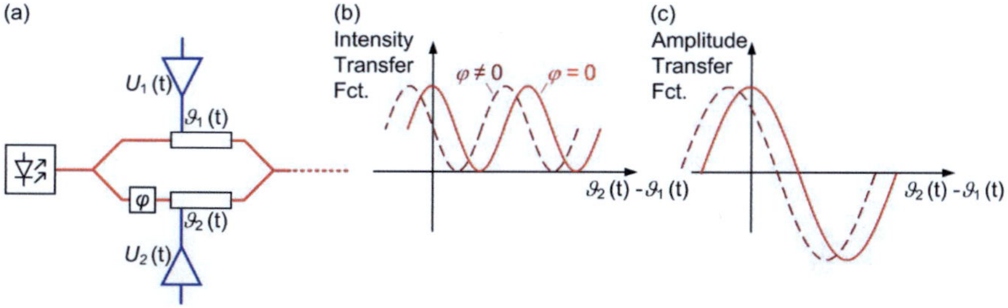

Figure 1.2 The configuration of an optical transmitter based on the Mach-Zehnder modulator (MZM). (a) The transmitter comprises a light source, electrical drivers and two high speed phase shifter built within the arms of the Mach-Zehnder interferometer [70].(b) The intensity and (c) the amplitude transfer functions of the single MZM.

Most of the silicon-based MZMs employ high speed phase modulators which use either the plasma dispersion effect in silicon (all-silicon MZMs) [36, 43–45, 72–75] or the Pockels effect in an electro-optic organic cladding. This technology is also known under the keyword "silicon organic hybrid (SOH)" [18, 22, 23]. The majority of the work done in developing MZMs was directed towards energy efficiency and small footprint [36, 43–45, 72–75]. A 750 µm long silicon MZM with an operating speed of 60 Gbit/s and an energy consumption of about 3.5 pJ/bit was demonstrated using the plasma dispersion effect [45]. Even lower energy consumption has been achieved with the SOH technology [23]. Devices fabricated in SOH technology benefit from the strong interaction of light with an electro-optic organic cladding, so that the required drive voltage is greatly reduced. Such phase modulators and MZMs can operate at speeds of 40 Gbit/s with an energy consumption of only 1.6 fJ/bit [76]. The modulation speeds can further be increased to 112 Gbit/s when using an IQ modulator and a 16 QAM format [22].

Typical voltage length product $U_\pi L$ are in the order one Vcm. For a low drive voltage, the device length has to be several hundreds of micrometers. Large dimensions of the MZMs make them costly and energy inefficient. Furthermore, all-silicon and SOH MZMs have limited RF bandwidth because of their large RC time constants and the mismatch between electrical and optical group velocities [77].

More compact MZMs have been achieved by using photonic crystal (PC) phase shifters (PC-MZM). A PC-MZM is demonstrated operating at 40 Gbit/s and having a length of 90 μm [19, 44, 78, 79]. However, the optical bandwidth limitations imposed by PCs make the PC-MZM not practical for communication systems utilizing wavelength diversity schemes. The performance of PC-MZMs is also extremely sensitive to fabrication errors and requires highly accurate lithography and precise etching.

In this thesis, we use plasmonics as an alternative path to realizing high-speed, compact and low-voltage phase modulators [J2, J3, C2, C3]. Using surface plasmon polaritons (gap SPPs) and the Pockels effect in electro-optic organic materials, we show that a plasmonic phase modulator (PPM) is feasible with a footprint of only $5 \times 6\ \mu m^2$ and with an RF bandwidth exceeding 65 GHz [J2]. The performance of our plasmonic modulator in comparison to state-of-the-art devices is given in the Table 1.1. It can be seen that our plasmonic phase modulator (PPM) not only offers the largest RF bandwidth, but also the smallest footprint with low electrical energy consumption comparable to very sensitive ring resonator modulators.

Table 1.1 State-of-the art silicon modulators compared to our plasmonic modulator.

	Photonic crystal [44]	Ring resonator [40]	Ring resonator [41]	Si-WG MZM [45]	SOH MZM [80]	This Work
Phase modulation	yes	no	no	yes	yes	yes
$U_\pi L$ [V cm]	NA			2	0.5	0.13
f_{3dB} [GHz]	NA	1	NA	27.8	22	> 65
Bit-rate [Gbit/s]	40	50	40	60	40	40
Pre-emphasis	no	yes	no	no	no	no
Operating range $\Delta\lambda$ [nm]	12	< 1	< 1	broad	broad	120
Footprint [μm^2]	~540	~150	~150	> 10^4	> 5×250	~ 29
Energy per bit @ 40Gbit/s [fJ/bit]	N. A.	N. A.	32 (+ FCS*)	3500	420	60
Insertion Loss [dB]	6.2	5.2	10.5	6.5	6	12

* Energy consumption of the feedback control system that is needed to stabilize the device temperature

2 Theoretical Background

Photonics and plasmonics are technologies which are concerned with generation, manipulation and detection of light [67]. In order to understand and improve the operation of photonic / plasmonic components such as waveguides, splitters, couplers and modulators, consideration of the wavelike nature of light is often sufficient. Here we review the classical electromagnetic (EM) wave theory according to [81, 82].

In this chapter, we first discuss the theory of classical electromagnetics in both dielectrics and metals. We discuss how the refractive index can be influenced by a second-order nonlinear effects which can be exploited to construct electro-optic modulators. The theory of surface plasmon polaritons (SPPs) is given in the last section of this chapter along with the descripttion of various SPP waveguides and the concept of the plasmonic phase modulator.

2.1 Theory of Electromagnetic Fields

Maxwell's four equations give a complete classical description of the electric and magnetic field interaction with matter. Maxwell's equations connect the electric displacement **D**, the magnetic induction **B**, the electric field **E** and the magnetic field **H**. The quantity ρ_{net} is the total net charge density, the **J** is the electric current density. In differential form we have [81]

$$\begin{aligned}
&\textit{Coulomb law} && \nabla\cdot\mathbf{D}(\mathbf{r},t)=\rho_{net}(\mathbf{r},t),\\
&\textit{Absence of free magnetic pole} && \nabla\cdot\mathbf{B}(\mathbf{r},t)=0,\\
&\textit{Faraday's law} && \nabla\times\mathbf{E}(\mathbf{r},t)=-\frac{\partial\mathbf{B}(\mathbf{r},t)}{\partial t},\\
&\textit{Ampere's law} && \nabla\times\mathbf{H}(\mathbf{r},t)=\frac{\partial\mathbf{D}(\mathbf{r},t)}{\partial t}+\mathbf{J}(\mathbf{r},t).
\end{aligned}\tag{2.1}$$

The displacement field **D** and the magnetic field **H** relate to the electric field **E** and magnetic inductance **B**, respectively, via the constitutive equations

$$\mathbf{D}(r,t)=\varepsilon_0\mathbf{E}(\mathbf{r},t)+\mathbf{P}(\mathbf{r},t),$$
$$\mathbf{H}(r,t)=\frac{\mathbf{B}(\mathbf{r},t)}{\mu_0}-\mathbf{M}(\mathbf{r},t), \tag{2.2}$$

where $\varepsilon_0 = 8.85419\times10^{-6}\,\mathrm{As/(Vm)}$ and $\mu_0 = 1.25664\times10^{-6}\,\mathrm{Vs/(Am)}$ are vacuum permittivity and permeability, respectively. The quantities **P** and **M** are macroscopic polarization and magnetization. For nonmagnetic materials the macroscopic magnetization **M** equals zero, which further simplifies the second equation of Eq. (2.2). The macroscopic polarization **P** is the sum of all dipole moments per volume of a material induced by an external stimulus **E**. The response of a spatially local material is described by a macroscopic polarization **P**, which in general has a nonlinear dependence on the electric field **E** [83]

$$\mathbf{P}(\mathbf{r},t)=\varepsilon_0\int_0^{+\infty}\underline{\chi}^{(1)}(\mathbf{r},t-t')\cdot\mathbf{E}(\mathbf{r},t')\mathrm{d}t'+$$
$$+\varepsilon_0\int_0^{+\infty}\int_0^{+\infty}\underline{\chi}^{(2)}(\mathbf{r},t-t',t-t''):\mathbf{E}(\mathbf{r},t')\mathbf{E}(\mathbf{r},t'')\mathrm{d}t'\mathrm{d}t''+\ldots, \tag{2.3}$$

where $\underline{\chi}^{(n)}$ is the nth order tensor of the influence function that defines the temporal memory of the material. The influence function is causal, $\underline{\chi}^{(n)}(\mathbf{r},t_1,t_2,\ldots \mathrm{t}_n)=0$ for $t_1,t_2,\ldots \mathrm{t}_n<0$, which is expressed by the lower integral boundaries in Eq. (2.3). The above expression between a polarization and an electrical field is valid for both isotropic and anisotropic materials as well as for dispersive and lossy media. It takes into account the fact that the polarization at time t is not only defined by the instantaneous electric field $\mathbf{E}(\mathbf{r},t)$, but also by the field at previous times. The integral in the first line of Eq. (2.3) describes the linear response of the material, while the following integrals describe the nonlinear parts of the response. The contributions of the nonlinear polarizations in the material response must be taken into account for strong electric fields.

Electric and magnetic fields wave equations can be obtained by eliminating either **E** or **H** in the Maxwell's equations. The wave equation for the electric field **E** results when taking the curl of both sides of Faraday's law,

$$\nabla\times\nabla\times\mathbf{E}(\mathbf{r},t)=-\mu_0\varepsilon_0\frac{\partial^2\mathbf{P}_L(\mathbf{r},t)}{\partial t^2}-\mu_0\varepsilon_0\frac{\partial^2\mathbf{P}_{NL}(\mathbf{r},t)}{\partial t^2}-\mu_0\frac{\partial\mathbf{J}(\mathbf{r},t)}{\partial t}-\mu_0\varepsilon_0\frac{\partial^2\mathbf{E}(\mathbf{r},t)}{\partial t^2}. \tag{2.4}$$

We have separated the linear $\mathbf{P}_{\mathrm{L}}$ and the nonlinear $\mathbf{P}_{\mathrm{NL}}$ terms of the polarization $\mathbf{P}$. As in our case the current density $\mathbf{J}$ is induced by the exciting electric field E, it can be regarded as a material property like the polarization $\mathbf{P}$.

2.1.1 Linear Optics

In linear optics, the strengths of electric fields interacting with matter are considered to be weak, and thus the nonlinear terms of the polarization are neglected resulting in a linear dependence of a displacement $\mathbf{D}$ on an electric field $\mathbf{E}$

$$\mathbf{D}(\mathbf{r},t)=\varepsilon_0\mathbf{E}(\mathbf{r},t)+\mathbf{P}_{\mathrm{L}}(\mathbf{r},t)=\varepsilon_0\mathbf{E}(\mathbf{r},t)+\varepsilon_0\int_0^{+\infty}\underline{\chi}^{(1)}(\mathbf{r},t-t')\cdot\mathbf{E}(\mathbf{r},t')\,\mathrm{d}t'. \tag{2.5}$$

The linear term of the polarization is denoted as $\mathbf{P}_{\mathrm{L}}$. We can replace the convolution integral in Eq. (2.5) with the more favorable multiplication operator by using a Fourier transform Eq. (A.1) of the physical quantities $\mathbf{D}$, $\mathbf{E}$ and $\mathbf{P}$, and by writing Eq. (2.5) in the frequency domain

$$\breve{\mathbf{D}}(\mathbf{r},\omega)=\varepsilon_0\breve{\mathbf{E}}(\mathbf{r},\omega)+\varepsilon_0\underline{\breve{\chi}}^{(1)}(\mathbf{r},\omega)\breve{\mathbf{E}}(\mathbf{r},\omega), \tag{2.6}$$

where ω is the angular frequency. The wave equation in the frequency domain can be derived from Eq. (2.4) for a linear medium, by assuming a linear dependence of electrical current density $\breve{\mathbf{J}}$ on the electric field $\breve{\mathbf{E}}$

$$\nabla\times\nabla\times\breve{\mathbf{E}}(\mathbf{r},\omega)==\omega^2\mu_0\varepsilon_0\breve{\mathbf{E}}(\mathbf{r},\omega)+\mu_0\varepsilon_0\omega^2\underline{\breve{\chi}}^{(1)}(\mathbf{r},\omega)\breve{\mathbf{E}}(r,\omega)-\mathrm{j}\mu_0\omega\underline{\breve{\sigma}}(\mathbf{r},\omega)\breve{\mathbf{E}}(\mathbf{r},\omega). \tag{2.7}$$

The quantity $\underline{\breve{\sigma}}(\mathbf{r},\omega)$ is the conductivity of a medium. The linear dependence of an induced current density on the external electric field gives a possibility to combine all the terms of the RHS and write the wave equation Eq. (2.7) in a simpler form [1]

$$\nabla \times \nabla \times \breve{\mathbf{E}}(\mathbf{r},\omega) - \frac{\omega^2}{c^2}\left(1 + \underline{\breve{\chi}}^{(1)}(\mathbf{r},\omega) - \mathrm{j}\frac{\underline{\breve{\sigma}}(\mathbf{r},\omega)}{\omega\varepsilon_0}\right)\breve{\mathbf{E}}(\mathbf{r},\omega) = 0. \tag{2.8}$$

We have defined the speed of light in vacuum as $c = 1/\sqrt{\mu_0\varepsilon_0}$. The modified form of the wave equation Eq. (2.8), which is valid for both non-conductive and conductive materials shows that no differentiation can mathematically be given between an imaginary part of the complex susceptibility $\underline{\breve{\chi}} = \underline{\breve{\chi}}' + \mathrm{j}\underline{\breve{\chi}}''$ and a conductivity $\underline{\breve{\sigma}}(r,\omega)$. Advanced ellipsometric techniques can measure the complex susceptibility at optical frequencies, and is often an easier measurement technique to implement than that of measuring the conductivity directly. Therefore, for the sake of convenience, we will assume zero conductivity throughout this thesis. Instead of regarding the conductivity, we will define a complex susceptibility whose imaginary part is responsible for all the absorption mechanisms, and which can be assessed easily by experiment and modeled in theory.

In optics, the linear susceptibility can be redefined as a relative permittivity or a refractive index to describe material properties, see Table 2.1. Even though all three descriptions are equivalent, the refractive index is mostly preferred, due to its direct connection to the wave velocity, see also Eq. (A.12),

$$\nabla \times \nabla \times \breve{\mathbf{E}}(\mathbf{r},\omega) - \frac{\omega^2}{c^2}\underline{n}^2(\mathbf{r},\omega)\breve{\mathbf{E}}(\mathbf{r},\omega) = 0, \tag{2.9}$$

where $\underline{n} = \underline{n}' - \mathrm{j}\underline{n}''$ is the complex refractive index tensor, whose real part $\underline{n}'$ is related to the phase velocity of light, and whose negative imaginary part $\underline{n}''$ is the extinction coefficient and determines the attenuation of a wave amplitude as shown in Eq. (A.17).

Table 2.1 Connection of linear susceptibility to the permittivity and the refractive index

Relative Permittivity	Refractive Index
$\underline{\varepsilon} = \underline{\varepsilon}' - \mathrm{j}\underline{\varepsilon}'' = 1 + \breve{\underline{\chi}}^{(1)}$	$\underline{n} = \underline{n}' - \mathrm{j}\underline{n}'' = \sqrt{1 + \breve{\underline{\chi}}^{(1)}}$

In the case of isotropic materials, the tensor quantities reduce to scalars, and the underscores are dropped.

2.1.2 Material Properties

Generally speaking the frequency dependence of the susceptibility is unknown and should be measured by, e. g., ellipsometry. However, for some isotropic materials the susceptibility $\breve{\chi}^{(1)}$ can be calculated by using simple classical Lorentz or Drude models [84]. The modelling of the susceptibility is important when a solution to the frequency dependent wave equation Eq. (2.8) should be found.

In the classical Lorentz model, the macroscopic response of a material is modelled assuming that it consists of a collection of microscopic harmonic oscillators representing the nuclei of an atom with an surrounding electron cloud [84]. In this case, the equation of motion of an electron in an electron cloud oscillation, under excitation by an external electric field can be given as

$$m_e \frac{\mathrm{d}^2 \mathbf{r}}{\mathrm{d}t^2} + m_e \gamma \frac{\mathrm{d}\mathbf{r}}{\mathrm{d}t} + m_e \omega_L^2 \mathbf{r} = -e\breve{\mathbf{E}}(\mathbf{r},\omega) e^{\mathrm{j}\omega t}. \tag{2.10}$$

The term on the LHS describes the drive force which oscillates with a frequency of ω. The quantity m_e is the electron mass, e is the value of electronic charge, ω_L and γ are the resonance frequency and the damping constant (collision frequency) of the oscillator, respectively. The second term on the LHS of Eq. (2.10) describes the energy loss mechanism while the third term is the elastic restoring force. The solution of the above equation of motion can be used to describe the microscopic dipole moments of individual atoms [85]

$$\breve{\mathbf{p}} = \frac{e^2 \breve{\mathbf{E}}}{m_e} \frac{1}{\omega_L^2 - \omega^2 + \mathrm{j}\gamma\omega}. \tag{2.11}$$

The macroscopic polarization is found by summing up all dipole moments [85]

$$\breve{\mathbf{P}}(\mathbf{r},\omega) = N\breve{\mathbf{p}} = \varepsilon_0 \breve{\chi}^{(1)}(\mathbf{r},\omega)\breve{\mathbf{E}}(\mathbf{r},\omega), \text{ where } \breve{\chi}^{(1)}(r,\omega) = \frac{Ne^2}{\varepsilon_0 m_e} \frac{1}{\omega_L^2 - \omega^2 + \mathrm{j}\gamma\omega}, \tag{2.12}$$

where N is the number of oscillators per volume and ε_0 is the vacuum permittivity. Eq. (2.12) gives the susceptibility of a material in the classical Lorentz model and defines the macroscopic linear polarization that is a result of the bound electron oscillating around a nucleus. The quantum mechanical analogy of Lorentz model are interband transitions of electrons between the energy levels with a separation close to $\hbar\omega_L$ [85]. To model the frequency dependence of the susceptibility over a large frequency range, the contribution of all of the oscillations with various resonance frequencies should be added together

$$\breve{\chi}^{(1)} = \sum_i \frac{N_i e^2}{\varepsilon_0 m_e} \frac{1}{\omega_{L_i}^2 - \omega^2 + \mathrm{j}\gamma_i\omega} \quad \text{with} \quad \sum_i N_i = N. \tag{2.13}$$

The parameter N_i is the number of electrons per volume that are bound to the nucleus with a resonance frequency of ω_{L_i}. Multiple resonances expressed in Eq. (2.13) describe the peaks in the absorption spectrum as well as the frequency dependence of the real part of the susceptibility, i.e. refractive index. Generally, the frequency dependence of the refractive index results in the spreading of optical pulses, therefore this effect is called dispersion.

The contribution of free electrons to the optical response of metals can be taken into account by the Drude model. In this case there is no restoring force (ω_L=0) in Eq. (2.10), and the complex susceptibility is

$$\breve{\chi}^{(1)} = -\frac{\omega_p^2}{\omega^2 - \mathrm{j}\gamma\omega}, \quad \text{and} \quad \varepsilon = \varepsilon_\infty - \frac{\omega_p^2}{\omega^2 - \mathrm{j}\gamma\omega}. \tag{2.14}$$

The parameter ε_∞ is the frequency independent background permittivity. The so-called plasma frequency ω_p is defined as

$$\omega_p^2 = \frac{Ne^2}{\varepsilon_0 m_e}. \tag{2.15}$$

The quantum mechanical analogy of the free electron response to an external electrical field is an electronic intraband transition [85].

When modelling the susceptibility of real metals at high photon energies, more often than not the Drude model is insufficient, and oscillations of the bound electrons in the inner electronic shells need to be considered as well [85]. This can be seen for instance in Figure 2.1(a), where the measured real part of the relative permittivity of silver is given in comparison to the values predicted by the Drude model (only free electrons) and by the Drude-Lorentz model (including bound electrons with three different resonance frequencies). It can be seen that the frequency dependence of the relative permittivity of silver can be precisely modelled in a wide frequency range using the Drude-Lorentz model. Nevertheless, the Drude model alone can still be considered to be a good approximation for susceptibilities of metals in the low-frequency range, see Figure 2.1(a). Figure 2.1(b) gives the refractive index and the extinction coefficient of silver as a function of the frequency ω. Metals show large extinction coefficients below their plasma frequency because of the large negative real part of the relative permittivity. Therefore, light in the frequency range below the plasma frequency does not propagate and has a very short penetration depth in into the metal.

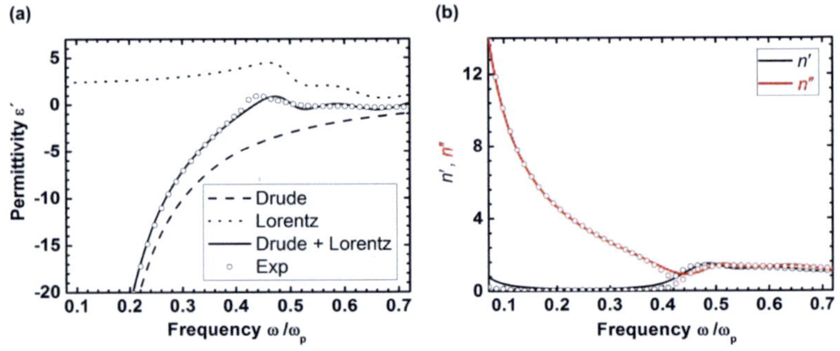

Figure 2.1 The relative permittivity and the refractive index of silver as a function of the optical frequency ω. (a) Experimental values of the relative permittivity Ref. [86] and the theoretical fits assuming the Drude model (free electrons only, dashed line) and the Drude-Lorentz (solid line, three resonance frequencies for bound electrons). The dotted line shows the contribution of the Lorentz model. (b) Measured refractive index (black circles) and extinction coefficient (red circles) of silver [86], and fit with a Drude-Lorentz model (solid lines).

In Appendix A.10, we give the optical properties and Drude parameters for silver, gold and copper. These noble metals are particularly interesting because of their relatively small electron collision frequencies (damping constant). Silver shows the least power dissipation at optical frequencies. Therefore, the insertion loss in plasmonic devices can be significantly reduced by using plasmonic silver waveguides. However, because silver oxidizes quickly, it is sometimes preferable to use gold despite of its large electron collision frequency [86], see Appendix A.10. Gold is known for its long-term stability which is important whenever the overall lifetime of a device matters most. Moreover, copper is widely used in CMOS technology, and therefore might be the only choice of metal available when designing mass-produced plasmonic ICs.

2.1.3 Nonlinear Optics and Second-Order Nonlinear Polarization

The nonlinear polarization in Eq. (2.3) must be taken into account either for large optical intensities, or for materials with strong nonlinear susceptibilities [87]. The interaction of light with a nonlinear medium is described with a nonlinear wave equation, which can be derived from Eq. (2.4),

$$\nabla \times \nabla \times \breve{\mathbf{E}}(\mathbf{r},\omega) - \frac{\omega^2}{c^2} \underline{n}^2(\mathbf{r},\omega)\breve{\mathbf{E}}(\mathbf{r},\omega) = \mu_0 \omega^2 \breve{\mathbf{P}}_{\mathrm{NL}}(\mathbf{r},\omega). \tag{2.16}$$

It can be seen that the nonlinear polarization acts as a secondary source giving rise to new spectral components in the EM wave spectrum [82]. The second-order nonlinear polarization term is only [82]

$$\mathbf{P}^{(2)}(\mathbf{r},t) = \varepsilon_0 \int_0^{+\infty} \int_0^{+\infty} \underline{\chi}^{(2)}(\mathbf{r},t-t',t-t'') : \mathbf{E}(\mathbf{r},t')\mathbf{E}(\mathbf{r},t'')\,\mathrm{d}t'\mathrm{d}t''. \tag{2.17}$$

The second-order polarization in the frequency domain is simply [83]

$$\breve{\mathbf{P}}^{(2)}(\mathbf{r},\omega) = \sum_{n,m} \varepsilon_0 \underline{\breve{\chi}}^{(2)}(\mathbf{r},\omega:\omega_n,\omega_m) : \breve{\mathbf{E}}(\mathbf{r},\omega_n)\breve{\mathbf{E}}(\mathbf{r},\omega_m). \tag{2.18}$$

The quantities $\breve{\mathbf{E}}(r,\omega_n)$ and $\breve{\mathbf{E}}(r,\omega_m)$ are the complex amplitudes (phasors) of the incident EM wave at frequencies ω_n and ω_m, respectively, and $\breve{\mathbf{P}}^{(2)}(r,\omega)$ is the Fourier transform of the second-order nonlinear polarization at $\omega = \omega_n + \omega_m$. The summation in Eq. (2.18) extends over all positive and negative pairs of frequencies (ω_n, ω_m) which lead to the same ω.

We assume now that the incident EM wave is a superposition of two monochromatic plane waves with frequencies of ω_1 and ω_2

$$\mathbf{E}(\mathbf{r},t) = \frac{1}{2}\left(\breve{\mathbf{E}}_1(\mathbf{r},\omega_1)e^{\mathrm{j}\omega_1 t} + c.c.\right) + \frac{1}{2}\left(\breve{\mathbf{E}}_2(\mathbf{r},\omega_2)e^{\mathrm{j}\omega_2 t} + c.c.\right). \tag{2.19}$$

Here, the positive and negative frequency components generate nonlinear polarization terms oscillating with various frequencies determined by all permutations of the incident frequencies for the special case $\omega_1 \geq \omega_2$, Table 2.2.

Table 2.2 Frequency components of the second-order nonlinear polarization for two monochromatic incident plane waves with $\omega_1 \geq \omega_2$.

SHG	SFG	DFG	EOR
$\omega_1 + \omega_1 = 2\omega_1$ $\omega_2 + \omega_2 = 2\omega_2$	$\omega_1 + \omega_2$	$\omega_1 - \omega_2$	$\omega_1 - \omega_1$ $\omega_2 - \omega_2$

SHG – second harmonic generation, DGF – difference frequency generation, SFG – sum frequency generation, EOR – electro-optic rectification.

When one of the fields in Eq. (2.19) is a static or low-frequency electric field with $\omega_2 = 0$, the second-order nonlinear polarization has a non-zero complex amplitude at the fundamental frequency of ω_1. Denoting the field strength of the low-frequency electric field as $\breve{\mathcal{E}}(\mathbf{r},0)$, we can write

$$\breve{\mathbf{P}}^{(2)}(\mathbf{r},\omega_1) = 2\varepsilon_0 \underline{\breve{\chi}}^{(2)}(\mathbf{r},\omega_1,0) : \breve{\mathbf{E}}_1(\mathbf{r},\omega_1)\breve{\mathcal{E}}(\mathbf{r},0). \tag{2.20}$$

Therefore, while solving the wave equation Eq. (2.16) for a fundamental frequency ω_1, the linear and second-order nonlinear polarizations can be combined on the left-hand side resulting in a wave equation similar to Eq. (2.9) valid for linear materials

$$\nabla \times \nabla \times \breve{\mathbf{E}}(\mathbf{r},\omega) - \frac{\omega^2}{c^2} \underline{n}_{NL}^2 \breve{\mathbf{E}}(\mathbf{r},\omega) = 0. \tag{2.21}$$

For simplicity, the frequency ω_1 is replaced by ω. It can be seen that the propagation of a wave in a non-centrosymmetric material can be described like wave propagation in a linear material with a refractive index $\underline{n}_{NL}$ which depends on the static electric field,

$$\underline{n}_{NL}^2(\mathbf{r},\omega) = \underline{n}^2(\mathbf{r},\omega) + 2\underline{\breve{\chi}}^{(2)}(\mathbf{r},\omega,0)\breve{\mathcal{E}}(\mathbf{r},0). \tag{2.22}$$

This effect, where the refractive index seen by an EM wave is varied by the presence of a static electric field is known as the Pockels effect or the linear electro-optic effect [82]. Many optical modulators employ the Pockels effect to modulate the phase of light, see Appendix A.3. In Chapter 4.1, we demonstrate a plasmonic phase modulator which uses Pockels effect to modulate the phase of the plasmonic signal.

Historically, an alternative mathematical formulation of the Pockels effect has been given using an electro-optic tensor r_{ijk} that describes the change of the index ellipsoid under the influence of the applied static electric field [82]. If the material is anisotropic and non-dispersive, the displacement field relates to the electric field via the relative permittivity tensor $\underline{\varepsilon}$,

$$\mathbf{D}(\mathbf{r},t)=\varepsilon_0\begin{pmatrix}\varepsilon_{xx} & \varepsilon_{xy} & \varepsilon_{xz}\\ \varepsilon_{yx} & \varepsilon_{yy} & \varepsilon_{yz}\\ \varepsilon_{zx} & \varepsilon_{zy} & \varepsilon_{zz}\end{pmatrix}\mathbf{E}(\mathbf{r},t). \tag{2.23}$$

For lossless and non-dispersive materials it can be shown that the permittivity tensor is a symmetric matrix with elements $\varepsilon_{ij}=\varepsilon_{ji}$ [82]. Moreover, the relative permittivity tensor can be converted to diagonal form by choosing an appropriate coordinate axes [88], which for linear materials results in

$$\left(\frac{1}{2\varepsilon_0 w_e}\right)\frac{D_x^2}{\varepsilon_{xx}}+\left(\frac{1}{2\varepsilon_0 w_e}\right)\frac{D_y^2}{\varepsilon_{yy}}+\left(\frac{1}{2\varepsilon_0 w_e}\right)\frac{D_z^2}{\varepsilon_{zz}}=1. \tag{2.24}$$

The quantity w_e is the electric field energy density which is constant for lossless materials [82]. The D_x, D_y, D_z are the Cartesian components of the displacement field in the newly chosen coordinate system, and $\varepsilon_{xx}, \varepsilon_{yy}, \varepsilon_{zz}$ are the diagonal coefficients of the transformed relative permittivity tensor. Eq. (2.24) describes an ellipse in its principle-axis system. Going back to an arbitrary Cartesian coordinate system $\tilde{x},\tilde{y},\tilde{z}$, the so-call index ellipsoid results [82],

$$\left(\frac{1}{n^2}\right)_{\tilde{x}\tilde{x}}\tilde{x}^2+\left(\frac{1}{n^2}\right)_{\tilde{y}\tilde{y}}\tilde{y}^2+\left(\frac{1}{n^2}\right)_{\tilde{z}\tilde{z}}\tilde{z}^2+2\left(\frac{1}{n^2}\right)_{\tilde{y}\tilde{z}}\tilde{y}\tilde{z}+2\left(\frac{1}{n^2}\right)_{\tilde{x}\tilde{z}}\tilde{x}\tilde{z}+2\left(\frac{1}{n^2}\right)_{\tilde{x}\tilde{y}}\tilde{x}\tilde{y}=1. \tag{2.25}$$

The quantities $(1/n^2)_{ij}$ are constants that describe the optical indicatrix in the new coordinate system [82]. The first-order change of the index ellipsoid due to the applied static electric field is linearly proportional to the applied field,

$$\left(\frac{1}{n^2}\right)'_{ij} = \left(\frac{1}{n^2}\right)_{ij} + \sum_k r_{ijk}\mathcal{E}_k \,. \tag{2.26}$$

The term on the RHS of Eq. (2.26) defines the optical indicatrix in the presence of the static electric field $\mathcal{E}_k$. The parameters r_{ijk} are known as electro-optic coefficients and give the rate at which the coefficients $(1/n^2)_{ij}$ change with the applied electric field. Similar to the permittivity tensor, the electro-optic tensor should be symmetric with respect to the indices ij [82]. This means that there are only 18 independent non-zero coefficients within the r_{ijk} tensor instead of full 27 coefficients. It allows us to construct the electro-optic matrix r_{sk} where ij indices are replaced with one index $s = \{1, 2, 3, 4, 5, 6\}$ and the index $k = \{\tilde{x}, \tilde{y}, \tilde{z}\}$ is redefined as $k = \{1, 2, 3\}$

$$ij = \begin{cases} \tilde{x}\tilde{x} \rightarrow s = 1 \\ \tilde{y}\tilde{y} \rightarrow s = 2 \\ \tilde{z}\tilde{z} \rightarrow s = 3 \\ \tilde{y}\tilde{z} = \tilde{z}\tilde{y} \rightarrow s = 4 \\ \tilde{x}\tilde{z} = \tilde{z}\tilde{x} \rightarrow s = 5 \\ \tilde{x}\tilde{y} = \tilde{y}\tilde{x} \rightarrow s = 6 \end{cases} \quad \text{and} \quad k = \begin{cases} \tilde{x} \rightarrow k = 1 \\ \tilde{y} \rightarrow k = 2 \\ \tilde{z} \rightarrow k = 3 \end{cases}. \tag{2.27}$$

If the material belongs to one of the crystallographic point groups, several coefficients of the matrix r_{sk} are zero, because the matrix r_{sk} must be invariant with the same symmetry operations that transform the crystal onto itself [89]. This then can simplify calculations.

Electro-optic (EO) organic materials are known for their large electro-optic coefficients [87]. They are primarily centro-symmetric because of the random distribution of dipole moments. Therefore, EO organic materials intrinsically do not show linear electro-optic effect. However, by breaking the centro-symmetricity of EO organic materials, via the so-called poling procedure, one can activate their electro-optic properties. The poling is normally done by applying a static electric field E_{pol} across the organic material at an elevated temperature [18, 87]. The poling field E_{pol} destroys the centro-symmetric nature of the EO organic material and creates an optical preference direction. Assuming that E_{pol} is oriented along the z-axis, we find the vanishing coefficients of the electro-optic matrix r by considering that it must be invariant with respect to a rotation with an arbitrary angle around the z-axis

$$\mathbf{r} = \begin{pmatrix} r_{11} & r_{12} & r_{13} \\ r_{21} & r_{22} & r_{23} \\ r_{31} & r_{32} & r_{33} \\ r_{41} & r_{42} & r_{43} \\ r_{51} & r_{52} & r_{53} \\ r_{61} & r_{62} & r_{63} \end{pmatrix} = \begin{pmatrix} 0 & 0 & r_{13} \\ 0 & 0 & r_{13} \\ 0 & 0 & r_{33} \\ 0 & r_{42} & 0 \\ r_{42} & 0 & 0 \\ 0 & 0 & 0 \end{pmatrix}. \tag{2.28}$$

Finally, the static electric field induced change in the index ellipsoid of an electro-optic organic material poled along the z-axis can be written as

$$\left(\frac{1}{n^2}\right)_{xx}\tilde{x}^2 + r_{13}\mathcal{E}_z\tilde{x}^2 + \left(\frac{1}{n^2}\right)_{yy}\tilde{y}^2 + r_{13}\mathcal{E}_z\tilde{y}^2 + \left(\frac{1}{n^2}\right)_{zz}\tilde{z}^2 + r_{33}\mathcal{E}_z\tilde{z}^2 + 2\left(\frac{1}{n^2}\right)_{yz} yz$$
$$+2r_{42}\mathcal{E}_y\tilde{y}\tilde{z} + 2\left(\frac{1}{n^2}\right)_{xz}\tilde{x}\tilde{z} + 2r_{42}\mathcal{E}_x\tilde{x}\tilde{z} + 2\left(\frac{1}{n^2}\right)_{xy}\tilde{x}\tilde{y} = 1. \tag{2.29}$$

Therefore, the static electric field $\mathcal{E}_z$ polarized along the poling direction z induces a refractive index change along all three axes. In case, if the incident optical wave has its dominant electric field component along the z-axis, the refractive index change Δn_{EO} seen by the wave can be given [87]

$$\Delta n_{\text{EO}} = -\frac{1}{2}\left(n_{\text{EO}}\right)^3 r_{33}\mathcal{E}_z. \tag{2.30}$$

We have assumed that the material is poled along the z-axis which resulted in a non-zero r_{33} coefficient and a refractive index change that can be described by Eq. (2.30). In case the poling direction is not along the z-axis the relation for the electro-optic tensor Eq. (2.28) does not hold. This is the case when later on discussing plasmonic phase modulators, where we assume a poling direction along x-axis. In this case, we use Eq. (2.30) to calculate the refractive index change along the x-axis while keeping the r_{33} notation for the electro-optic coefficient.

Plasma dispersion effect

The nonlinear polarization induced in a medium can also be a result of the nonlinear dependence of the electron concentration N on an electric field ε. This can for example happen in metals and semiconductors, where the charge density can be changed locally due to the electric field screening effect. Any locally induced charge density modifies the Drude parameter N, and therefore the susceptibility of a medium. In this case, Eq. (2.3) can be written as [90]

$$\breve{\mathbf{P}}(\mathbf{r},\omega)=\varepsilon_0\breve{\chi}^{(1)}\breve{\mathbf{E}}(\mathbf{r},\omega)+2\varepsilon_0\breve{\chi}^{(2)}\mathbf{E}(\mathbf{r},0)\breve{\mathbf{E}}(\mathbf{r},\omega)=-\varepsilon_0\frac{N\left(\mathbf{E}(\mathbf{r},0)\right)e^2/m_e\varepsilon_0}{\omega^2-\mathrm{j}\omega\gamma\left(\mathbf{E}(\mathbf{r},0)\right)}\breve{\mathbf{E}}(\mathbf{r},\omega). \tag{2.31}$$

Basically, the free carrier density N and the damping constant γ become functions of the static electric field $\mathbf{E}(\mathbf{r},0)$. However, we assume that the damping constant γ remains independent of $\mathbf{E}(\mathbf{r},0)$, because in metals and in semimetals, the main collision mechanism is primarily governed by electro-phonon scattering. The field-induced charge carrier density change ΔN results in a refractive index change Δn similar to Eq. (2.30),

$$\Delta n=-\frac{1}{2n}\frac{\Delta Ne^2/m_e\varepsilon_0}{\omega^2-\mathrm{j}\gamma\omega}. \tag{2.32}$$

The equation describes the plasma dispersion effect, where an electric field induced change of the charge carrier density modifies the refractive index. The effect is exploited widely in all-silicon modulators [11]. For an arbitrary conductor, the calculation of the refractive index change given by Eq. (2.32) requires an estimation of the carrier density change ΔN. In Chapter 4.2 we discuss a general method for calculating the induced charge carrier density change ΔN by solving the Poisson equation in Thomas-Fermi approximation.

2.1.4 Optical Waveguide Theory

Examples of waveguide geometries are given in Figure 2.2. Here we have defined the propagation direction along the z-axis. A strip waveguide comprises a high-refractive index dielectric core surrounded by a low refractive index substrate with a cladding material on top, see Figure 2.2 (a). Light is confined to the core by total internal reflection [91]. However, a slot waveguide as in Figure 2.2 (b) confines the light to the low-index region of the slot between the rails with refractive indices of n_{rl} [91]. In both cases, the refractive indices of cladding n_{cl} and substrate n_{sb} must be smaller than the refractive indices of the core n_{cr} and of the rails n_{rl}.

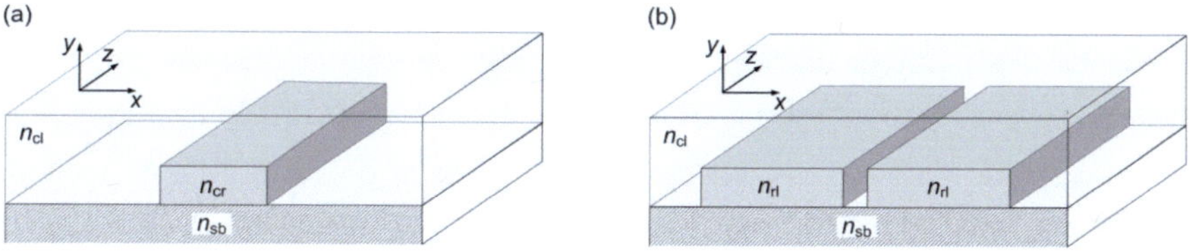

Figure 2.2 Examples of optical waveguide geometries. (a) Strip waveguide comprising a high refractive index n_{cr} core surrounded by a low index substrate and cladding materials with refractive indices of n_{sb} and n_{cl}, respectively. Light is guided along the z-axis confined in a core material due to the total internal reflection. (b) Slot waveguide comprising two rails with refractive indices n_{rl}. Light is confined to the low-index region inside the slot.

Fields in a longitudinally invariant waveguide can be divided into two groups. The first group represents the waves that are guided along the waveguide while maintaining their transverse field profiles. These waves are called guided modes. The second group includes the waves that describe a power radiation away from the waveguide, and are called radiating modes [92],

$$\begin{aligned}\breve{\mathbf{E}}(\mathbf{r},\omega) &= \sum_{\mu} \breve{\mathbf{e}}_{\mu}(x,y,\omega)e^{-\mathrm{j}\beta_{\mu}z} + \breve{\mathbf{e}}_{\text{rad}}(\mathbf{r},\omega),\\ \breve{\mathbf{H}}(\mathbf{r},\omega) &= \sum_{\mu} \breve{\mathbf{h}}_{\mu}(x,y,\omega)e^{-\mathrm{j}\beta_{\mu}z} + \breve{\mathbf{h}}_{\text{rad}}(\mathbf{r},\omega).\end{aligned} \tag{2.33}$$

The quantities $\breve{\mathbf{e}}_\mu(x,y,\omega)$ and $\breve{\mathbf{h}}_\mu(x,y,\omega)$ are the field structures of the μth guided mode of a waveguide and can be found by solving the corresponding eigenvalue problem [93]. The quantity β_μ is the propagation constant of the μth mode. Given the frequency ω, the values of β_μ are determined from an eigenvalue problem [92]. The resulting so-called dispersion relation $\beta_\mu(\omega)$ includes both the material dispersion and the waveguide dispersion. The parameter β_μ is often specified relative to the vacuum wave vector $k_0 = \omega/c$ using an effective refractive index n_{eff},

$$\beta = k_0 n_{\text{eff}}. \tag{2.34}$$

Different modes are characterized by different propagation constants β_μ which determines the phase velocity v_p of the μth mode, see Eq. (A.13) in Appendix A.3. If a waveguide sustains only one propagating mode, it is said to be singlemoded, otherwise it is multi-moded. In general, the transverse components of the electric $\breve{\mathbf{e}}_\mu(x,y,\omega)$ and the magnetic $\breve{\mathbf{h}}_\mu(x,y,\omega)$ field structures are complex. However, for a guided mode in a lossless dielectric waveguide they are real [92]. Thus, based on the discussion in Appendix A.2, the Cartesian components of the fields profiles $\breve{\mathbf{e}}_\mu(x,y,\omega)$ and $\breve{\mathbf{h}}_\mu(x,y,\omega)$ are in phase.

In the general case, the electric and magnetic field profiles $\breve{\mathbf{e}}_\mu(x,y,\omega)$ and $\breve{\mathbf{h}}_\mu(x,y,\omega)$ have non-zero components along all three axes in a Cartesian coordinate system. Nevertheless, while the polarization of the modes varies across the cross-section, a mode can still be characterized by its dominant polarization. A mode is said to be quasi-TE if it has a dominant electric field component parallel to the substrate, i. e., along the x-direction, see Figure 2.3(a). Similarly, a quasi-TM mode has a dominant electric field that is perpendicular to the substrate plane as can be seen in Figure 2.3(b). Figure 2.3(c) gives the effective refractive index as a function of the wavelength for a silicon strip waveguide with a height of 220 nm and a width of 450 nm. Because of the asymmetry of the silicon waveguide geometry, the quasi-TE mode is confined stronger to the silicon core than the quasi-TM mode. Consequently, the quasi-TE mode has effective index larger than the one of the quasi-TM mode.

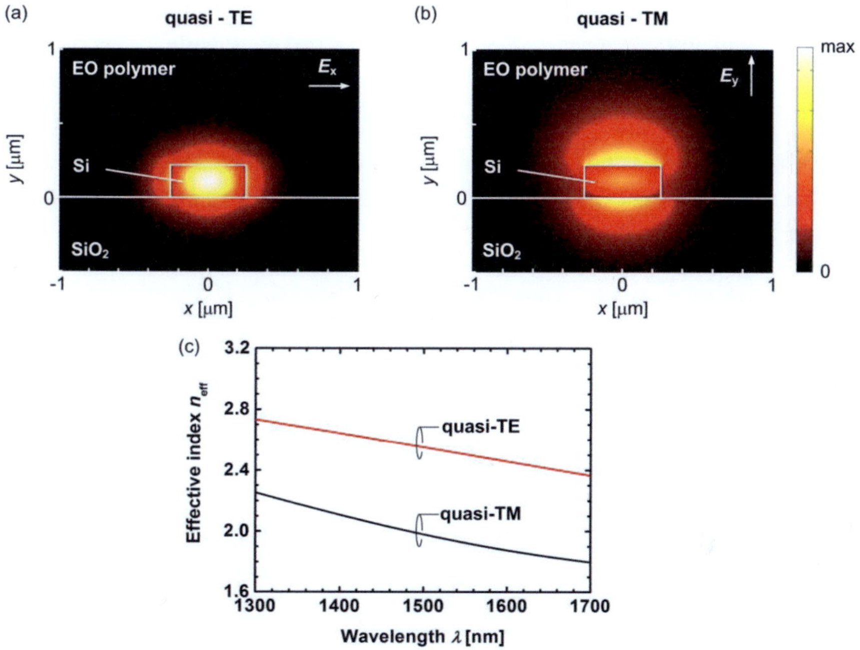

Figure 2.3 Fundamental quasi-TE and quasi-TM modes of a silicon strip waveguide with an EO organic cladding. (a) Field profiles of a quasi-TE (a) and a quasi-TM (b) modes and corresponding effective refractive indices (c).

2.2 Introduction to Surface Plasmon Polaritons

Surface electromagnetic waves exist at the interface of materials which have opposite signs of their respective permittivity [1]. It has been discussed in Section 2.1.2 that metals show a negative relative permittivity below their plasma frequency. Thus, in the same frequency range, metal-insulator interfaces sustain electromagnetic surface waves, so called surface plasmon polaritons (SPPs) [2].

2.2.1 Single Metal-insulator Interface

The basic structure sustaining a SPP is a single metal-insulator interface given in Figure 2.4. Here, the SPP dispersion relation can easily be derived using the wave equation in the frequency domain, Eq. (A.7), and considering a slab configuration with metal (Layer 1) and insulator (Layer 2) filling the half spaces $x<0$ and $x>0$, respectively [1]. Fields $h_y^{(1)-}$ and $h_y^{(2)+}$ of the guided SPPs decay exponentially along positive and negative x-directions. Therefore, the ansatz for the electric and magnetic fields of transverse magnetic wave (TM wave) can be written for each half space as [1]

$$\left.\begin{aligned} h_y(x,z,\omega) &= h_y^{(1)-}(\omega)e^{\mathrm{j}k_1x}e^{-\mathrm{j}\beta z} \\ e_{x,z}(x,z,\omega) &= e_{x,z}^{(1)-}(\omega)e^{\mathrm{j}k_1x}e^{-\mathrm{j}\beta z} \end{aligned}\right\} \text{ for } x<0,$$
$$\left.\begin{aligned} h_y(x,z,\omega) &= h_y^{(2)+}(\omega)e^{-\mathrm{j}k_2x}e^{-\mathrm{j}\beta z} \\ e_{x,z}(x,z,\omega) &= e_{x,z}^{(2)+}(\omega)e^{-\mathrm{j}k_2x}e^{-\mathrm{j}\beta z} \end{aligned}\right\} \text{ for } x>0. \quad (2.35)$$

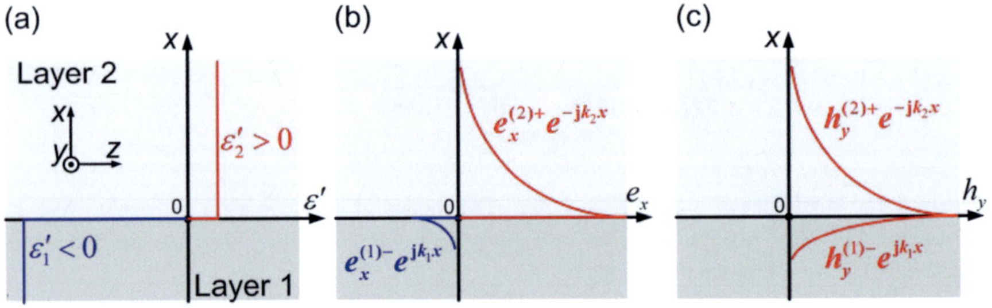

Figure 2.4 Metal-insulator interface supporting a SPP. (a) Permittivity, (b) electric and (c) magnetic field distributions across the interface.

We assumed that the wave propagates along the positive z-direction with a complex propagation constant of β whose real part gives the phase evolution and the imaginary part the amplitude decay along z, see Appendix A.2. The complex quantities k_1 and k_2 are the magnitudes of the transverse wave vectors along the x-axis. In the ansatz of Eq. (2.35) we have assumed that only the y-component of the magnetic field is non-zero, which means that the y-component of the electric field must be zero based on the Faraday's law. The relation between propagation constant β and transverse wave vectors $k_{1,2}$ can be found by substituting the ansatz Eq. (2.35) of the magnetic field into the wave Eq. (A.7) for each half space,

$$k_{1,2} = k'_{1,2} - \mathrm{j}k''_{1,2} = \sqrt{k_0^2 \varepsilon_{1,2} - \beta^2} \, . \tag{2.36}$$

The quantity $k_0 = \omega/c$ is the wave vector of light in vacuum. Relative permittivity is defined as ε, propagation constant β as well as transverse wave numbers are complex quantities . Using boundary conditions for the transverse magnetic $h_y^{1-} = h_y^{2+}$ and the longitudinal electric $e_z^{1-} = e_z^{2+}$ fields, the SPP dispersion relation can be derived [1]

$$\beta = \beta' - \mathrm{j}\beta'' = k_0 \sqrt{\frac{\varepsilon_1 \varepsilon_2}{\varepsilon_1 + \varepsilon_2}} \, . \tag{2.37}$$

The SPP dispersion relations for a silver-silica and a silver-silicon interface are plotted in Figure 2.5. Here, the frequency dependence of the relative permittivity of silver is taken into account by the Drude model, see Section 2.1.2. For frequencies up to the so-called surface plasmon frequency ω_{sp}, the real part of the SPP propagation constant is larger than the propagation constant of a hypothetical plane wave in the insulator layer adjacent to the metal, see Figure 2.5(a) and Figure 2.5(b) [1]. Consequently, the transverse components k_1 and k_2 of the wave number have large negative imaginary parts k_1'' and k_2''. Bound SPPs are confined to the metal-insulator interface and have exponentially decaying fields both along the positive and the negative x-direction. Therefore, the transverse components of the wave numbers k_1 and k_2 have to be chosen such that their imaginary parts are negative, see Figure 2.5(c) and Figure 2.5(d). The penetration depth $1/k_i''$ describes the field decay in material i. For frequencies $\omega < \omega_{\mathrm{sp}}$, the penetration depth in metal is much shorter than that in the insulator. Moreover, the penetration depth in the insulator decreases significantly when increasing the carrier frequency, while the penetration depth in the metal stays relatively constant up to $0.4 \times \omega_{\mathrm{p}}$, see Figure 2.5(c). Increasing the relative permittivity of the insulator increases the SPP propagation constant for a given frequency as well as it reduces the frequency range up to ω_{sp} corresponding to the guided SPPs, see Figure 2.5(a) for a silver-silica interface and Figure 2.5(b) for a silver-silicon interface. When approaching the surface plasmon frequency ω_{sp}, the SPP attenuation constant increases strongly. The increase of the SPP propagation loss at frequencies close to ω_{sp} can be attributed to the reduction of the SPP group velocity as can be seen in Figure 2.5(a) and Figure 2.5(b). For real-world lossy conductors the transverse wave numbers k_1 and k_2 have non-zero real parts, see Figure 2.5(e) and Figure 2.5(f). This indicates that there is a power flow also in the transverse x-direction. Inside the insulator ($x > 0$), the transverse component of the Poynting vector is directed along the negative x-direction for frequencies up to ω_{sp}, whereas inside the metal the transverse wave number, therefore, the Poynting vector changes sign from negative to positive at low frequencies, see lower left inset of Figure 2.5(e).

Figure 2.5 SPP dispersion diagrams for a single metal-insulator interface. The normalizing plasma frequency is $\omega_p = 2\pi\times2190\times10^{12}\,s^{-1}$. The minimum frequency for the curves is $\omega = 2\pi\times100\times10^{12}\,s^{-1}$. Left column: Silver-silica interface. Right column: Silver-silicon interface. (a),(b) Longitudinal SPP wave numbers. Real part β' (black), negative imaginary part β'' (red). Light lines in green. (c),(d) Negative imaginary parts of the transverse SPP wave numbers (decay constants) in metal k_1'' (red) and in insulator k_2'' (black). The frequency ω_{sp} defines the upper cutoff frequency until which bound SPPs exist. The larger the insulator's permittivity ε_2, the narrower the frequency range of bound SPPs. Well below ω_{sp} the decay constant in metal is significantly larger than in the insulator layer. For frequencies close to ω_{sp}, SPPs are strongly confined to the metal-insulator interface with similar decay constants in both metal and insulator layers. (e),(f) Real parts of transverse SPP wave numbers in metal k_1' (red) and in insulator k_2' (black). Light lines in green. A positive transverse wave number k_1' in the metal (inset, $x < 0$) shows that for small frequencies the power flow is directed away from the interface into the metal. With larger frequencies, the sign of k_1' is negative, and the power flow is directed towards the dielectric.

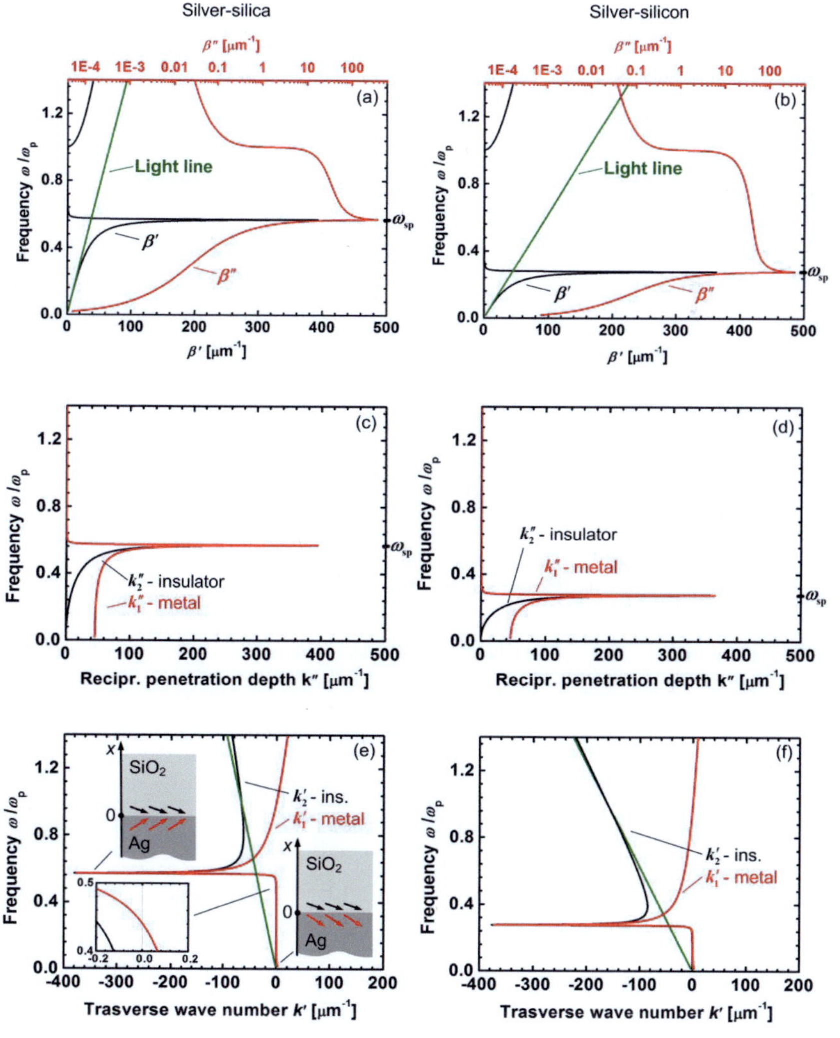
Silver-silica
Silver-silicon
β″ [μm⁻¹]
1E-4 1E-3 0.01 0.1 1 10 100
(a)
(b)
(c)
(d)
(e)
(f)
Frequency ω / ω_p
Light line
β′
β″
ω_sp
β′ [μm⁻¹]
k″₂ - insulator
k″₁ - metal
Recipr. penetration depth k″ [μm⁻¹]
SiO₂
Ag
x
k′₂ - ins.
k′₁ - metal
Trasverse wave number k′ [μm⁻¹]

2.2.2 Metal-insulator-metal Configuration

A better confinement of SPPs can be achieved by using more advanced plasmonic structures comprising more than one metal-insulator interfaces. In this case, the SPPs propagating along the various interfaces couple to each other, which leads to a degeneracy of the dispersion diagram. Metal-insulator-metal (MIM) and insulator-metal-insulator (IMI) structures are examples of such plasmonic waveguides [94, 95]. Both for MIM and IMI configurations, the SPPs of two metal-insulator interfaces result in two guided SPP modes with symmetric and anti-symmetric field structures [94, 95]. In the case, SPPs are classified to be either symmetric or anti-symmetric based on the sign of the longitudinal components of their electric fields and on the charge distribution at the metal surfaces [96]. This differs from the definition usually used in other fields of photonics, where modes are classified either symmetric or anti-symmetric based on a transverse component of the electric field. In the case of MIM and IMI structures, the dispersion relation can be numerically calculated by applying the matrix method as described in Appendix A.9.

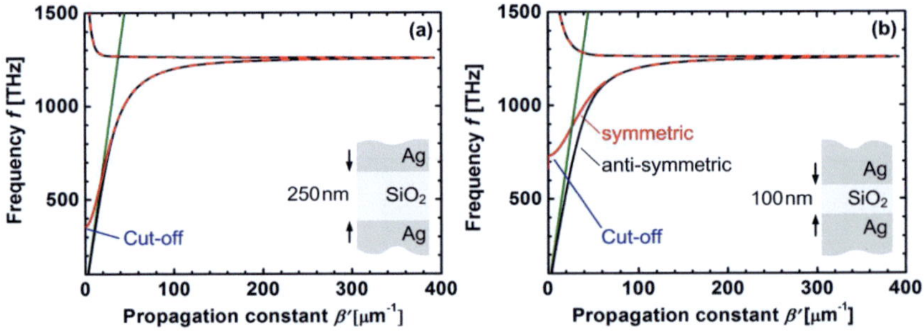

Figure 2.6 Dispersion relation of silver (Ag) - silicon dioxide (SiO_2) - silver (Ag) MIM waveguide. Dispersion relation of both symmetric and anti-symmetric SPPs for an insulator thicknesses of 250 nm (a) and 100 nm (b). The light line in SiO_2 is given as a green solid line

The dispersion relations of symmetric and anti-symmetric SPPs for an Ag - SiO_2 - Ag structure with a SiO_2 thicknesses of 250 nm and 100 nm are given in Figure 2.6(a) and Figure 2.6(b), respectively. The symmetric SPP exhibits a cut-off frequency which is defined by the material properties and by the thickness of an insulator layer sandwiched between two metals. Figure 2.6 shows that the lower cut-off frequency increases with decreasing insulator thickness. Unlike the symmetric SPP, the anti-symmetric mode does not exhibit a cut-off frequency and covers the entire frequency range of the single-interface SPP as shown in Figure 2.5. The real part of the electric and magnetic field profiles for a frequency of 900 THz are given in Figure 2.7. Both symmetric (red) and anti-symmetric (black) SPPs modes are plotted for the Ag - SiO_2 - Ag structure with a SiO_2 thickness of 100 nm. The electric field of an anti-symmetric mode is strongly confined to the insulator layer as can be seen from the transverse fields in Figure 2.7(middle row).

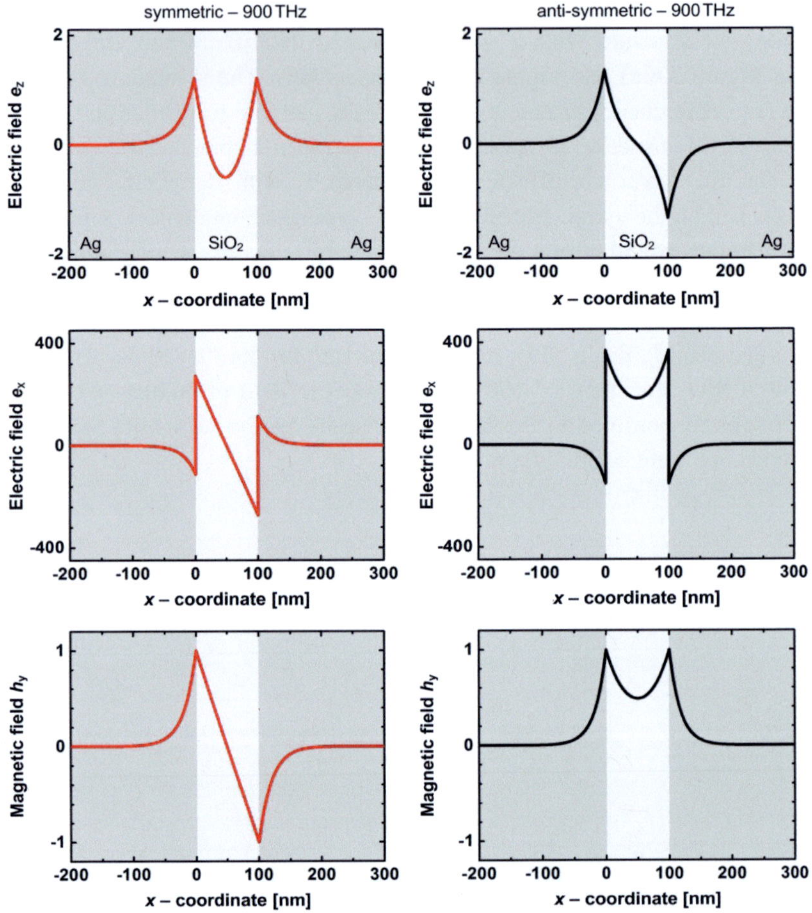

Figure 2.7 Symmetric and anti-symmetric SPPs of Ag-SiO_2-Ag waveguide with a SiO_2 thickness of 100 nm. Non-zero components of the electric and magnetic transverse fields (middle and lower row) are plotted for the symmetric SPP (red solid line, left column) and for the anti-symmetric SPP (right column, black solid line). The upper row shows the longitudinal symmetric and anti-symmetric electric fields.

For applications in the infrared range, the anti-symmetric mode is in particular interesting because it shows a strong confinement of the field to the insulator layer for all operating frequencies and all insulator thicknesses. The effective refractive index and the group refractive index are given in Figure 2.8 as a function of the wavelength for various insulator refractive indices. It can be seen that the SPP phase velocity, which is defined by the effective refractive index n_{eff}, reacts sensitively to the refractive index of an insulator sandwiched between the metals. This fact is used later on for building plasmonic modulators. Moreover, the sensitivity of the SPP phase velocity when changing the refractive index of an insulator increases for shorter wavelengths because of the reducing group velocity, Figure 2.8(b). In general, the insulator thickness and the insulator refractive index are important parameters that can be used to tune phase and group velocity of an SPP.

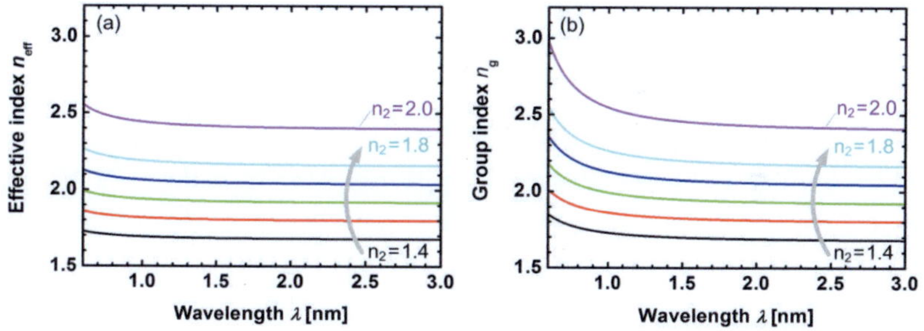

Figure 2.8 Effective and group refractive indices for an Ag-SiO_2-Ag slab waveguide with a 100 nm thick SiO_2 layer. (a) Effective refractive index and (b) group refractive index as a function of the wavelength for various insulator refractive indices from 1.4 to 2.0.

2.2.3 Surface Plasmon Polariton Waveguides

The three-dimensional (3D) analogue of a metal-insulator-metal slab waveguide is a metallic slot waveguide (MSW), see Figure 2.9 [94, 96–100]. Two configurations of MSWs have been reported in the literature with either vertical or horizontal slot orientations, Figure 2.9 [99]. Here we assumed that the substrate is made of silicon dioxide (SiO_2).

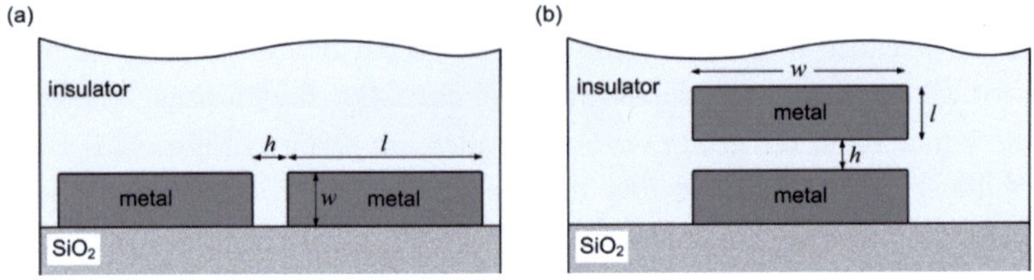

Figure 2.9 Cross-sections of 3D metallic slot waveguides (MSW) with (a) vertically orientated (vertical MSW) and (b) horizontally oriented slots (horizontal MSW).

In the previous section, we showed that the slot size h is a geometrical parameter that can be used to tune the properties of SPPs of the MIM slab structure. For real three-dimensional MSWs such as in Figure 2.9, the geometry of the metal defines the SPP dispersion diagram. In particular, metals dimensions w and l influence the effective and group refractive indices. The propagation constant and the field profiles of SPP modes of both vertical and horizontal MSWs can be numerically found using the software package based COMSOL Multiphysics which employs a the finite-element-method, see Appendix A.8.

A vertical MSW comprises two metal rails separated by a distance of several tens of nanometers, see Figure 2.9(a). The slot between the rails is filled with an insulator which is often chosen to be the same as the entire cladding material on top of the substrate. Unlike the high-index dielectric strip waveguide discussed in Section 2.1.4, the MSW cladding can have an arbitrary refractive index. Multiple metal-insulator interfaces within a MSW lead to various SPP modes with symmetric and anti-symmetric charge distributions similar to the MIM slab waveguides. Figure 2.10 gives the transverse components of the electric fields of four lowest-order SPP modes in a vertical MSW. The electric SPP fields in a vertical MSW have non-zero components along all three coordinate axes. Thus, unlike a MIM slab waveguide, the modes can only be classified as quasi-TM or quasi-TE, similar to the dielectric strip waveguide discussed in Section 2.1.4. The anti-symmetric SPP, a so-called gap SPP, is shown in Figure 2.10(Mode 4). The gap SPP has the highest effective refractive index among all guided SPPs and shows the highest confinement to the slot.

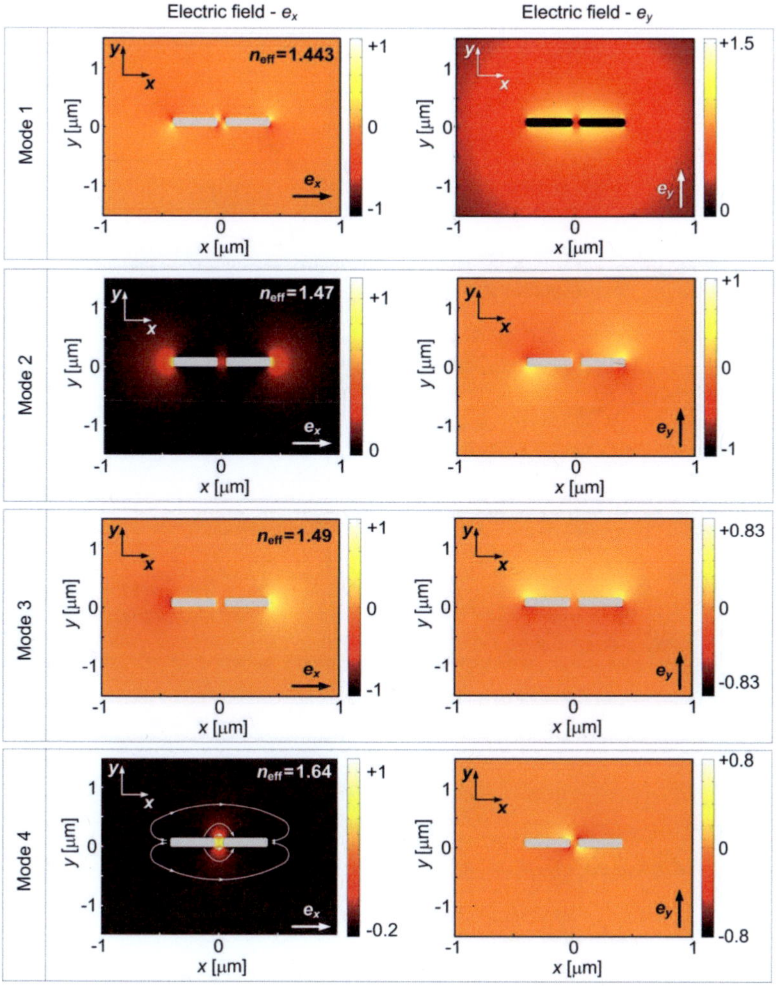

Figure 2.10 SPP modes of a vertical metallic slot waveguide (vertical MSW) with a slot width $h = 140$ nm and a metal width $l = 750$ nm, ordered according to the effective refractive indices (fundamental mode at bottom row).The metal thickness w is chosen to be 150 nm. The metals's complex refractive index is $0.55 + j11.1$ Silica with a refractive index of 1.44 is taken both as cladding and substrate material. SPP modes are given in terms of the transverse components of the electric field e_x and e_y. Both components e_x and e_y are normalized by the maximum value of e_x.

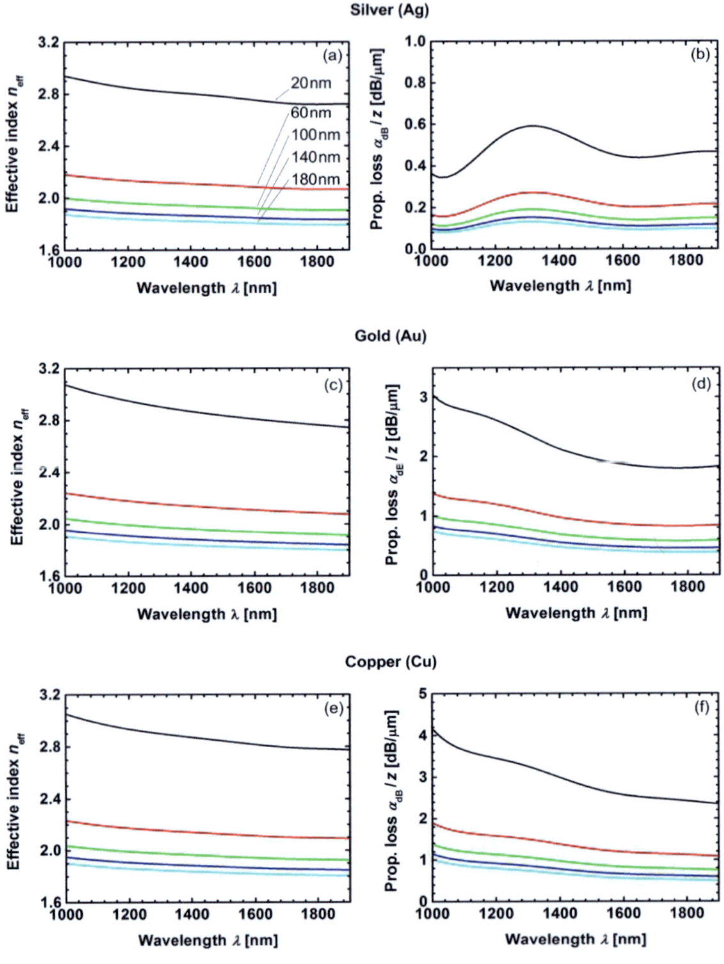

Figure 2.11 Calculated effective refractive indices n_{eff} and power propagation loss α_{dB} of the gap SPPs in vertical MSWs made of silver, gold and copper. The metal properties were measured [86] and interpolated for drawing the graphs. (a), (c), (e) Effective refractive index of a gap SPP as a function of wavelength for various slot widths h of 20 nm (black), 60 nm (red), 100 nm (green), 140 nm (blue), and 180 nm (cyan). (b),(d),(f) Power propagation loss as a function of the operating wavelength for slot widths of 20 nm (black), 60 nm (red), 100 nm (green), 140 nm (blue), and 180 nm (cyan). It can be concluded that silver has the smallest propagation loss by far.

We plot the effective refractive index and the power propagation loss α_{dB} of a gap SPP as a function of wavelength for three different metals and for five different slot widths h of 20 nm (black), 60 nm (red), 100 nm (green), 140 nm (blue), and 180 nm (cyan). We consider silver, gold and copper because of their low optical losses. Wavelength dependence of the measured complex refractive indices of silver, gold and copper are fitted with a polynomial as described in Appendix A.10. Fitted values of the complex refractive indices are then used in the simulations. The MSW is assumed to be covered with an electro-optic organic material M3 with a refractive index of 1.68. For this example, we choose the metal thickness $w = 150$ nm and the rail width $l = 2$ µm. Similar to the slab MIM waveguide, the effective refractive index increases when reducing the wavelength λ and the slot size h, see Figure 2.11. For a fixed wavelength, the reduction of the slot size increases the gap SPP dispersion. Consequently, SPP group velocity reduces similar to the case of an MIM as discussed previously. However, this then increases the power loss of a gap SPP similar to the case of a single interface as discussed in Section 2.2.1. Figure 2.11 also gives the power loss α_{dB} Eq. (A.18) as a function of the operating wavelength λ. The expected loss of 0.2 ...3 dB / µm is typical for a gap SPP at a wavelength of 1550 nm and makes its use in the integrated optics and in telecommunications really challenging. The power loss α_{dB} of the gap SPP in a silver MSW with a slot size $h = 100$ nm can be as low as 0.2 dB / µm. A power loss of about 1 dB / µm is estimated for MSWs made of both gold and copper layers.

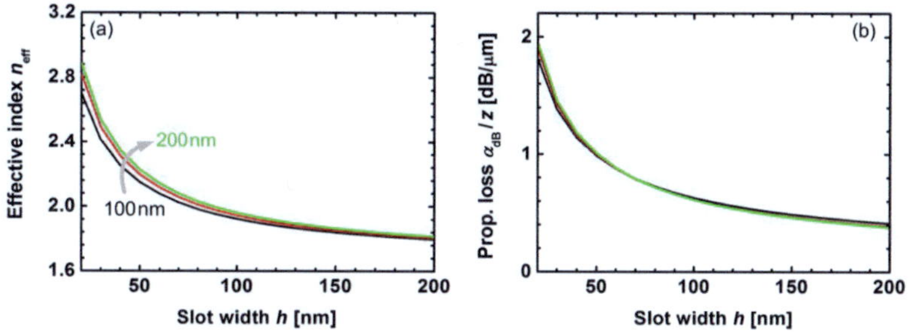

Figure 2.12 (a) Effective refractive index n_{eff} and (b) power loss α_{dB} of a gap SPP at a wavelength of 1550 nm as function of slot width h. The results are given for three different metal thicknesses of 100 nm (black), 150 nm (red), and 200 nm (green).

The effective refractive index and the power loss of a gap SPP are given in Figure 2.12 as a function of the slot size h. To understand the influence of the finite extension w of the metal layers, we consider three different metal thicknesses of $w = 100$ nm (in black), $w = 150$ nm (red), and $w = 200$ nm (green). The effective refractive index of the gap-SPP abruptly increases when reducing the slot size h, as can be seen in Figure 2.12. The length of a plasmonic device employing a MSW with the slot width of 70 … 150 nm should preferably not exceed 30 … 40 µm in order to be still practical. Devices longer than 40 µm are expected to suffer from extremely high power loss of about 16 … 28 dB. The effective refractive index n_{eff} and the attenuation constant α_{dB} exhibit only a slight dependency on the metal thickness w, see Figure 2.12.

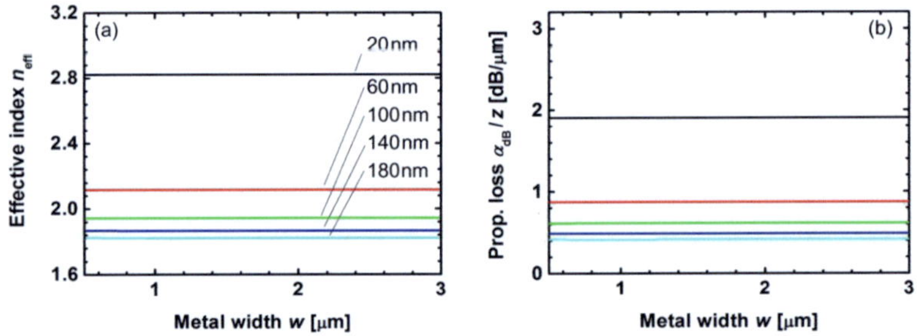

Figure 2.13 (a) Effective refractive index and (b) attenuation constant α_{dB} of the gap SPP as a function of the metal width l for five different slot sizes of 20 nm (black), 60 nm (red), 100 nm (green), 140 nm (blue) and 180 nm (cyan).

Figure 2.13 gives the effective refractive index n_{eff} and the power loss α_{dB} of a gap SPP as function of the metal width l. The parameter l plays only a negligible role in defining the dispersion diagram of the SPP whenever it is $l \leq 0.5$ µm holds. The metal width l can, however, influence the confinement of the gap SPP to the insulator slot whenever it is made smaller than 0.5 µm. The minimum acceptable metal width l defines the overall footprint $2 \times l \times L$ of a device, where L is the device length.

2.2.4 Plasmonic Phase Modulator

It has been previously discussed that the propagation constant and the power loss of a gap SPP strongly depend on the refractive index of the insulator which fills the slot. For instance, filling the slot with an electro-optic material discussed in Section 2.1.3, the phase of the gap SPP can be modulated by an applied static electric field. A device where the phase of an SPP is modulated by an electrical signal is called a plasmonic phase modulator (PPM). Applying the first-order perturbation theory discussed in Appendix A.7 [19], the phase modulation $\Delta\phi$ achieved in a PPM can be calculated for a MSW with a length of L and a slot size of h by

$$\Delta\phi = -\Delta\beta L = k_0 \frac{1}{2}\frac{UL}{h} r_{33} n_{\mathrm{EO}}^3 \Gamma, \tag{2.38}$$

where $\Delta\beta$ is the change in a propagation constant of a gap SPP due to the small refractive index perturbation $\Delta n_{\mathrm{EO}} = -U r_{33} n_{\mathrm{EO}}^3 / 2h$, U is the voltage between the metals and $k_0 = \omega / c$ is the wave vector of light in vacuum. We define an interaction factor Γ as

$$\Gamma = \frac{2n_{\mathrm{EO}}}{Z_0} \frac{\int\limits_{\text{Active area}} \breve{\mathbf{e}}_x(x,y,\omega)\breve{\mathbf{e}}_x(x,y,\omega)\,\mathrm{d}x\mathrm{d}y}{\int\limits_M \left[\breve{\mathbf{e}}(x,y,\omega)\times\breve{\mathbf{h}}(x,y,\omega)+\breve{\mathbf{e}}(x,y,\omega)\times\breve{\mathbf{h}}(x,y,\omega)\right]_z \mathrm{d}x\mathrm{d}y}. \tag{2.39}$$

The quantity $Z_0 = \sqrt{\mu_0/\varepsilon_0}$ is the vacuum wave impedance. The interaction factor Γ is a parameter which describes how strong the light-matter interaction is in a PPM. It is determined by the geometry of the PPM, by material properties of the metal and of the nonlinear material as well as by the operating wavelength. An important parameter for such a phase modulator is the voltage U_π which results in a phase change of π

$$U_\pi L = \frac{2\pi h}{k_0 r_{33} n_{\mathrm{EO}}^3 \Gamma}. \tag{2.40}$$

As a quality measure, the voltage-length product $U_\pi L$ is useful: The smaller the $U_\pi L$ product is, the better the performance of the device. A phase modulator with a small $U_\pi L$ product has a small footprint (small L) and consumes little energy (small U). In Figure 2.14, we give the values of $U_\pi L$ products that can be achieved in a plasmonic phase modulator with vertical configuration, filled with a nonlinear material with an electro-optic coefficient of $r_{33} = 70\mathrm{pm/V}$, see Appendix A.10. The $U_\pi L$ values as a function of the operating wavelength are displayed for five different slot widths and for three different metals, namely silver, gold and copper. Similar $U_\pi L$ values are estimated for all three metals. The $U_\pi L$ decreases when reducing the operating wavelength and the slot size h, see Figure 2.14. This is a consequence of the reduction of the SPP group velocity as already discussed in the previous sections. A product $U_\pi L < 100\,\mu\mathrm{m\,V}$ can be achieved with MSWs having sub-50 nm slot widths. However, one should note that reducing the operating wavelength and the slot size can dramatically increase the propagation loss of the gap SPP as discussed in Section 2.2.3. Thus, a more meaningful quality metric (figure–of–merit, FOM) refers the effective SPP propagation length L_e to the $U_\pi L$ product (see Appendix A.2),

$$\mathrm{FOM} = \frac{L_e}{U_\pi L}. \tag{2.41}$$

A device with a large FOM combines small optical loss with a small footprint and a small electric energy consumption. It can be seen that silver provides the largest FOM because of its small power attenuation. It is important to note that the device performance is better for short operating wavelengths despite the resulting large propagation loss. Moreover, the FOM increases with reducing the slot size h. This is a result of the fact that by decreasing the slot size h the $U_\pi L$ product decreases faster than the effective propagation length L_e.

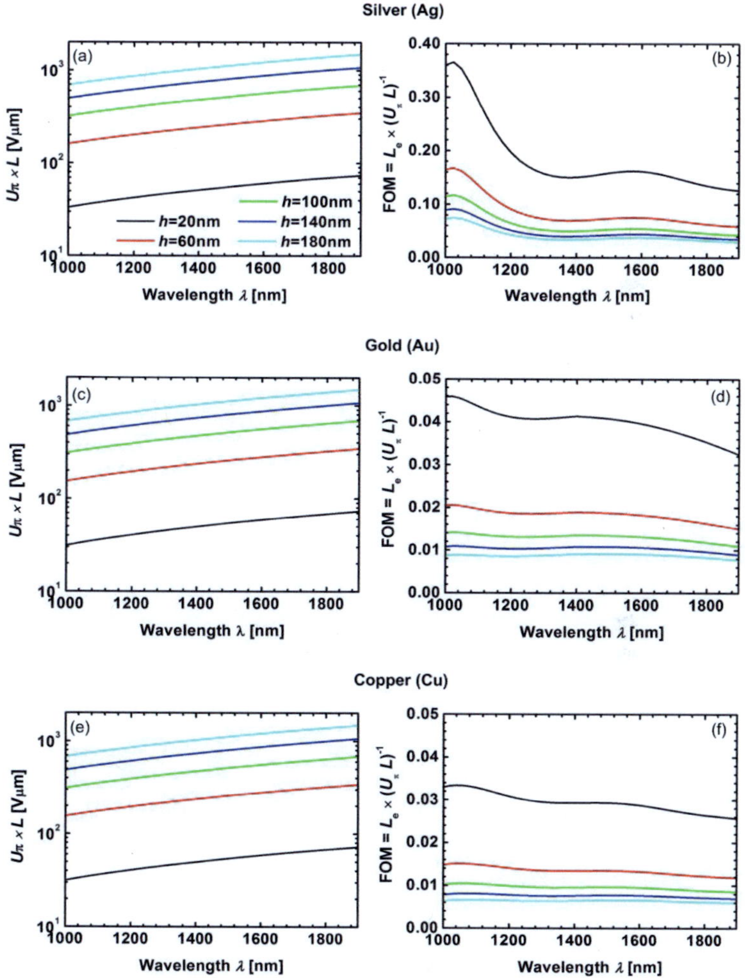

Figure 2.14 Characteristics of plasmonic phase modulators (PPMs) utilizing an electro-optic material with $r_{33} = 70$ pm/V. The $U_\pi \times L$ product and the $\mathrm{FOM} = L_e/(U_\pi L)$ are given for PPMs made of silver (a), gold (b) and copper (c) electrodes. The best device performance can be achieved in a PPM made of silver.

3 Passive Plasmonic Components

Merging plasmonics with silicon photonics requires efficient converters between the elements fabricated in different technologies. Particularly, to use gap SPPs for optical modulation, efficient methods are needed to convert a photonic mode of a silicon strip waveguide into a gap SPP [101–115].

In this chapter, we introduce two mode conversion schemes for efficient conversion of the photonic modes of a silicon strip waveguide into the gap SPPs of MSWs in both vertical and horizontal configuration. In addition, we demonstrate how the strong polarization sensitivity of a MSW can be used for building polarization beam splitters for silicon photonics.

3.1 Photonic-to-plasmonic Mode Converter: TM-modes

The content of this section is a direct copy of the Journal publication [J1].

A. Melikyan, M. Kohl, M. Sommer, C. Koos, W. Freude, and J. Leuthold, "Photonic-to-plasmonic mode converter," Opt. Lett. 39, 3488-3491 (2014).

Minor changes have been done to adjust the position of figures and notations of variables.

[Start of Paper]

Abstract: A novel photonic-to-plasmonic mode converter for efficiently converting a silicon strip waveguide mode to a gap surface plasmon polariton (SPP) of a metallic slot structure is proposed. A conversion efficiency of more than 85 % is found for metallic slots with a slot size of 30–50 nm. Calculations show that high conversion efficiencies can be achieved for various cladding materials with refractive indices of 1.44, 1.6 and 1.7. The optical operating bandwidth of the converter is around 40 nm. The proposed mode converter shows a good tolerance with respect to fabrication errors, and it requires a simple fabrication procedure only.

3.1.1 Introduction

Surface plasmon polaritons (SPP) are key to a new generation of plasmonic integrated circuits (ICs) with dimensions that are compatible with electronic IC sizes [2, 116]. A bottleneck of plasmonics is, however, the high optical loss originating in metals. A typical 0.5 dB / μm propagation loss of a SPP [117] confines their application range to ultra-short components only. To solve this issue, the combination of compact plasmonic active components such as modulators, switches, phase shifters and lasers with other low-loss silicon photonic strip waveguides (WG) seems to be attractive and might lead to new ICs comprising SPP devices, photonic circuits and electronics [118]. The merging of plasmonics and low-loss silicon photonics, however, calls for an efficient photonic-to-plasmonic mode converters.

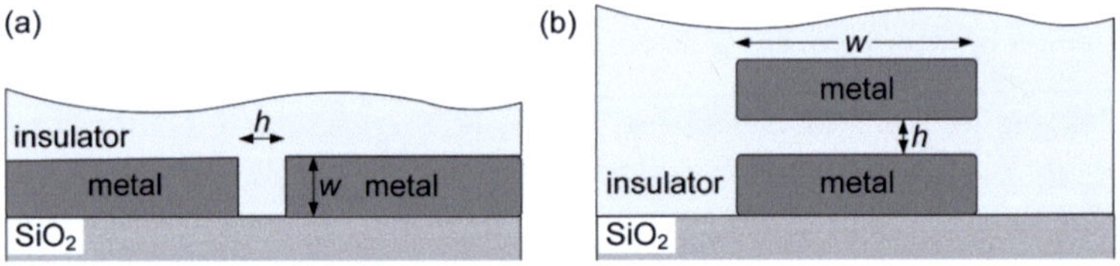

Figure 3.1 Metallic slot waveguides in two different configurations. (a) Vertically oriented and (b) horizontally oriented metallic slot waveguides.

A basic plasmonic building block which can easily be cointegrated with electronic ICs is the metallic slot waveguide (MSW) [94, 97–99, 119]. MSWs are not just plasmonic waveguides but can be used as, e. g., plasmonic photodetectors, light sources and modulators. The latter for example can be achieved by filling the slot with an active material, such as an electro-optic organic material [24, 25, 120–122], a semiconductor [37, 38] or a metal oxide [54, 55, 68, 123], and by applying a modulation voltage. Two configurations of MSWs have been reported [99] with either vertically [94, 97–99, 119] or horizontally [100] orienttated slots as shown in Figure 3.1(a) and Figure 3.1(b), respectively. When MSW structures are used as modulators, it is of particular important that the separation *h* between the metals is designed to be small, i.e., below 100 nm, and that the width *w* is larger than 150 nm for achieving a strong field enhancement in the slot. However, fabrication of a metallic slot in Figure 3.1(a) with a large *w* / *h* aspect ratio as well as excitation of its gap SPP becomes challenging. While several methods have been reported to excite the gap SPP in vertical MSWs [101, 102, 104], fabrication of high aspect ratio slots in a vertical configuration is still a major issue. Alternatively, the horizontal MSWs of Figure 3.1(b) can more easily be fabricated with a large aspect ratio by means of a bottom-up approach. However, the proposed approaches to excite the gap SPP in the horizontal MSW require a highly accurate lithography step and do not provide easy control over the coupling efficiency [100].

In this letter, we propose a novel coupling scheme between a silicon strip waveguide and a horizontal metallic slot waveguide with a conversion efficiency of more than 85 % over a broad band of optical frequencies. Particularly, we show that a photonic mode of the silicon strip waveguide with a height of 340 nm and a width of 300 nm can be fully converted into a gap SPP of a sub-50 nm metallic slot filled with an insulator material having refractive indices of 1.44, 1.6 and 1.7. We also discuss the influence of various fabrication errors on the coupler performance. The proposed converter can be used in various plasmonic devices where an efficient and controllable excitation and extraction of the gap SPP is needed.

3.1.2 Design and Analysis

The operation principle of the proposed mode converter is based on similar ideas as underlying the multimode interference coupler (MMI), see Figure 3.2 [124]. A photonic mode propagating through a silicon strip waveguide excites supermodes in the photonic/plasmonic MMI-based coupling section (CS) of length *L*. By properly selecting the length *L* and other geometrical parameters, a gap SPP can efficiently be excited in the end of the CS. Depending on the application, the silicon waveguide can either be terminated after the CS or extended beyond this point. Here, for the sake of the example we select a silicon strip waveguide with a height of 340 nm and a width of 300 nm.

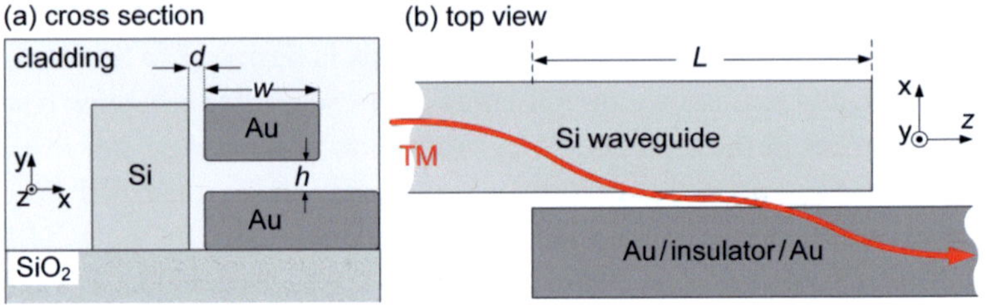

Figure 3.2 (a) Cross-section and (b) top view of the suggested coupler. The TM mode launched into the silicon strip waveguide couples to a plasmonic waveguide via a coupling section of length L.

We optimize the device's performance employing the simplified eigenmode expansion (EME) [125] method. To find the coupling strength of arbitrary eigenmodes i and j, we use the general orthogonality relation, see also Appendix A.5 (replacing ν with i, μ with j) [92]

$$\underbrace{\frac{1}{4}\iint\left[\breve{\mathbf{e}}_i\times\breve{\mathbf{h}}_j+\breve{\mathbf{h}}_i\times\breve{\mathbf{e}}_j\right]_z \mathrm{d}x\mathrm{d}y}_{C^{ij}}=\underbrace{\delta_{ij}\frac{1}{2}\iint\left[\breve{\mathbf{e}}_i\times\breve{\mathbf{h}}_j\right]_z \mathrm{d}x\mathrm{d}y}_{P_{ij}}. \tag{3.1}$$

Thus, the normalized overlap integral between modes i and j can be given by Eq. (A.31) by replacing in with j, ν with i

$$c_i^j=\frac{C^{ij}}{\sqrt{P_{ii}P_{jj}}}. \tag{3.2}$$

Neglecting backward propagating waves, we study light propagation in the CS as a superposition of all supermodes, including guiding, leaky and radiating modes. The distributions of the electric $\breve{\mathbf{E}}_{\mathrm{CS}}(\mathbf{r},\omega)$ and the magnetic fields $\breve{\mathbf{H}}_{\mathrm{CS}}(\mathbf{r},\omega)$ in the CS can be written as

$$\breve{\mathbf{E}}_{\mathrm{CS}}(\mathbf{r},\omega)=\sum_{\mu=1}^{N} c_\mu^s \breve{\mathbf{e}}_\mu(x,y,\omega)\cdot e^{-\mathrm{j}\beta_\mu z},$$
$$\breve{\mathbf{H}}_{\mathrm{CS}}(\mathbf{r},\omega)=\sum_{\mu=1}^{N} c_\mu^s \breve{\mathbf{h}}_\mu(x,y,\omega)\cdot e^{-\mathrm{j}\beta_\mu z}. \tag{3.3}$$

Here, the indices μ and s stand for supermode and silicon strip waveguide photonic modes, respectively. The quantities $\breve{\mathbf{e}}_\mu(x,y,\omega)$ and $\breve{\mathbf{h}}_\mu(x,y,\omega)$ are the electric and the magnetic fields of the μth supermode. The propagation constant and the excitation efficiency of the μth supermode are defined as β_μ and c_μ^s, respectively. The excitation efficiency c_μ^s is calculated as an overlap integral Eq. (3.2) between the silicon waveguide mode and a specific supermode μ^{th}.

The supermodes of the CS are simulated with a finite element method (FEM), see Appendix A.8 and [126]. To estimate the final photonic-to-plasmonic mode conversion efficiency, first the gap SPP is independently simulated for a homogenous cladding, and then the distributions of the electric $\breve{\mathbf{e}}_{\mathrm{MSW}}(x,y,\omega)$ and the magnetic field $\breve{\mathbf{h}}_{\mathrm{MSW}}(x,y,\omega)$ are calculated. The final conversion efficiency is computed as a normalized overlap integral $c_{\mathrm{MSW}}^{\mathrm{CS}}(z)$ of the gap SPP with the field distributions at the CS $\breve{\mathbf{E}}_{\mathrm{CS}}(\mathbf{r},\omega)$ and $\breve{\mathbf{H}}_{\mathrm{CS}}(\mathbf{r},\omega)$.

Figure 3.3(b) shows the magnitude of the overlap integral $\left|c_{\mathrm{MSW}}^{\mathrm{CS}}(z)\right|$ versus the z coordinate for a distance d of 20 nm (red) and 60 nm (black). The maximum value of $\left|c_{\mathrm{MSW}}^{\mathrm{CS}}(z)\right|$ is taken as the best conversion efficiency, and the corresponding z-coordinate is the respective optimum CS length L. As can be seen, the distance d mainly influences the coupling length L. The larger the distance d, the longer the coupling length L becomes.

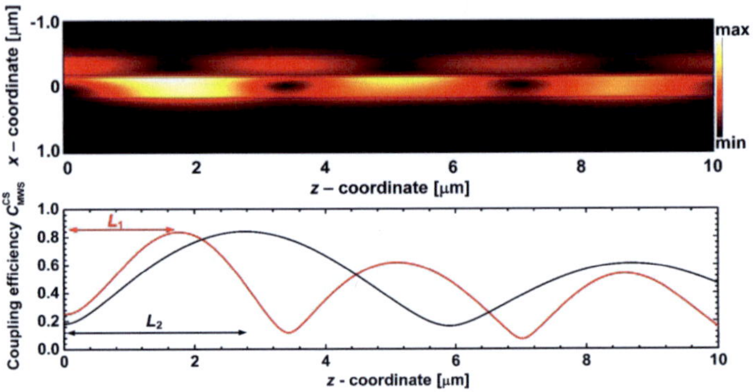

Figure 3.3 Electrical field distribution $\left|\breve{\mathbf{E}}_{\mathrm{CS}}(\mathbf{r},\omega)\right|_y$ in the mode converter with a width w of 300 nm, slot height h of 60 nm is given for an organic cladding with a refractive index of 1.7. The conversion efficiency is given as a function of a length L for two different distances d of 20 nm (red) and 60 nm (black).

3.1.3 Results and Discussion

The conversion efficiency $c^{\mathrm{CS}}_{\mathrm{MSW}}$ and the coupling length are given in Figure 3.4 as a function of the distance d and the slot height h for MSWs with a width w of 200 nm and for various cladding materials. Glass with a refractive index of 1.44, see Figure 3.4(a), (b), and organic materials with refractive indices of 1.6, see Figure 3.4(c), (d), and 1.7, see Figure 3.4(e), (f), are considered. For all three types of cladding materials, conversion efficiencies exceeding 85 % can be achieved for MSWs with sub-50 nm slots. Moreover, by varying geometrical parameters, e.g., the distance d, the conversion efficiency can be tuned. This might be needed e. g. in the case of plasmonic lasers. The proposed coupling scheme shows great tolerance with respect to fabrication errors, which impair the target length L. Additionally, the variation in conversion efficiency, associated with errors of ±20 nm in defining the distance d, is below 5%.

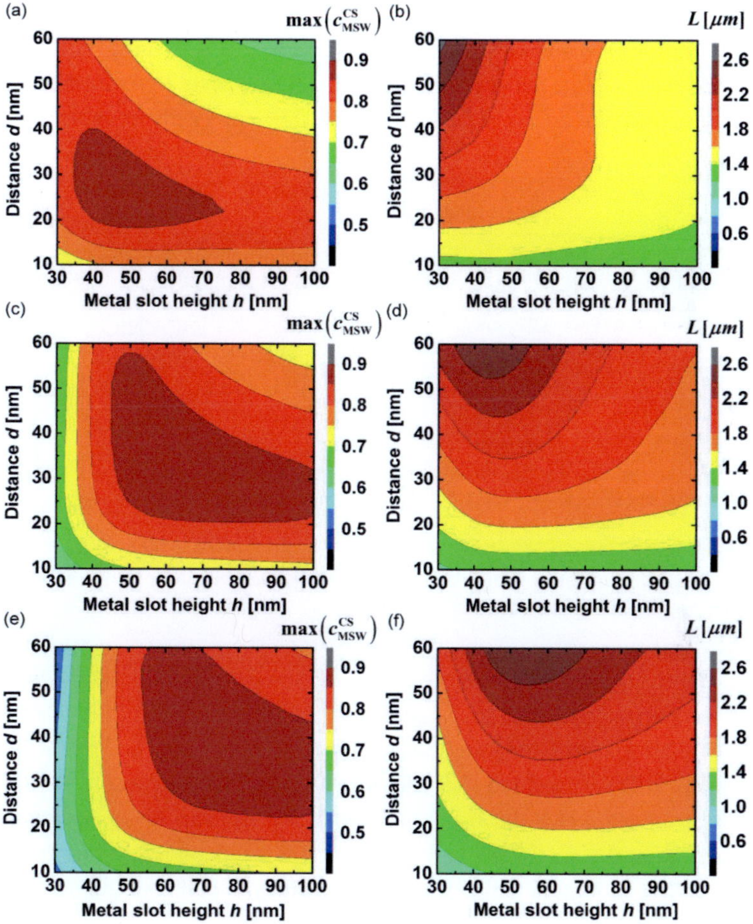

Figure 3.4 (a), (c), (e) Conversion efficiency $c_{\mathrm{MSW}}^{\mathrm{CS}}$ and (b), (d), (f) optimum CS length L for a MSW with a width w of 200 nm and for various cladding materials: (a) and (b) for glass, (c), (d) and (e), (f) for organic materials with n of 1.6 and 1.7 respectively.

Furthermore, we study the influence of the MSW width w on the device performance. The conversion efficiency and the CS length as calculated for a 300 nm wide MSW with a cladding refractive index of 1.7 are given in Figure 3.5(a) and

Figure 3.5(b), respectively. A conversion efficiency of 80 % can be achieved for a metallic slot with a height of 50 nm, i.e., for a MSW with an aspect ratio of 6. MSW with such a large aspect ratio are difficult to realize in a vertical configuration, whereas in the horizontal configuration they can easily be fabricated.

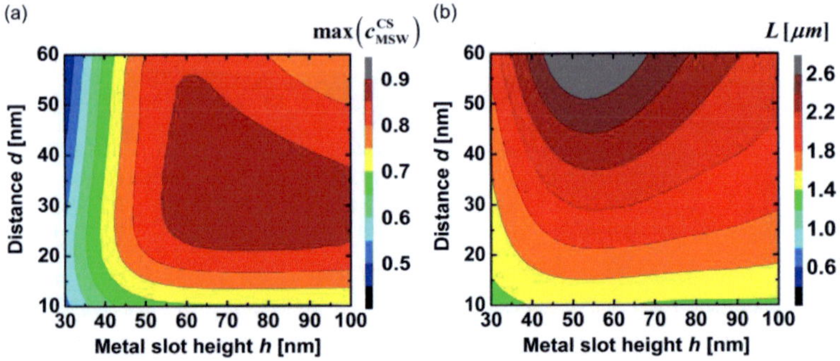

Figure 3.5 Conversion efficiency and coupling length for a MSW with a width w of 300 nm and a cladding refractive index of 1.7. (a) Conversion efficiency and (b) coupling length for various distances d and slot sizes h.

To investigate how the performance of the proposed mode converter depends on the carrier wavelength, we investigate the mode conversion mechanism in a converter with a silicon dioxide cladding numerically using a finite-difference time-domain (FDTD) method [127]. In this particular simulation, a metallic slot height h of 30 nm and a distance d of 60 nm are chosen. A continuous silicon waveguide is used with a CS length of $2 \times L$ of 6.8 μm, as plotted in the inset of Figure 3.6. Transmission and reflection spectra are given in Figure 3.6. The wavelength dependence of the optical properties of Au is taken into account by an appropriate Drude model. No strong resonance behavior is seen in the optical response of the mode converter. The operating wavelength region of the proposed device is in the range of 40 nm, which is comparable to the one reported for metallic tapered mode converters.

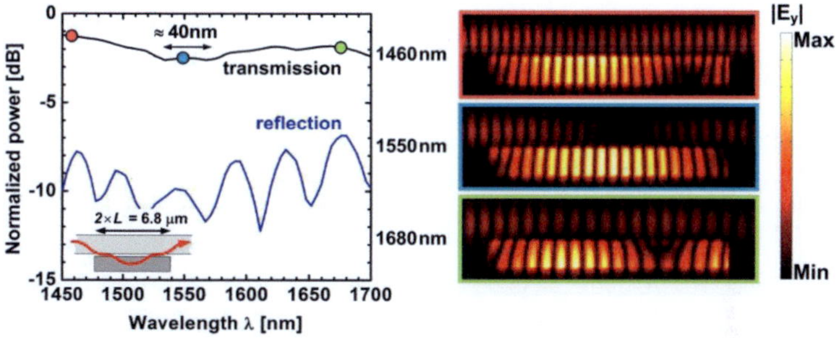

Figure 3.6 Transmission and reflection spectra in the converter with a height h of 30 nm, a distance d of 60 nm and a cladding material with a refractive index of 1.44. In addition, the electric field distributions for three different carrier wavelengths are given in the graph on the right-hand side. It can be seen that the photonic mode is fully converted into a gap SPP for a wavelength of 1550 nm.

3.1.4 Conclusion

To conclude, we report on a novel photonic-to-plasmonic mode converter for silicon photonics. The approach provides more than 85 % conversion efficiency for a gap SPP in a horizontal metallic slot waveguide, where the sub-50 nm slot is filled with various insulator materials. The proposed mode converter does not exhibit any resonant behavior and can be operated in a wide wavelength range. It requires only a simple "bottom-up" fabrication approach and shows good tolerance with respect to fabrication errors.

[End of Paper]

3.2 Photonic-to-plasmonic Mode Converter: TE-modes

Light contained in the mode of a dielectric waveguide can be converted into a gap SPP in a vertical MSW efficiently by using metallic tapered mode converters (MTMC) [101, 105–107, 109–111].

In this section, we present the design of an MTMC that can be used to efficiently convert the quasi-TE mode of a silicon strip waveguide into a gap SPP of the vertical MSW filled with an electro-optic organic material with a refractive index of 1.68. We discuss the performance of an optimized MTMC and present the characterization results of the fabricated converters.

3.2.1 Design and Analysis

A metallic tapered mode converter comprises a tapered dielectric waveguide with a certain tapering angle of θ and two metal electrodes located in both sides of the waveguide, see Figure 3.7 (a). Light guided through the dielectric waveguide launches the SPP hybrid modes at the tapered section which is then adiabatically guided into the MSW. We use MTMCs for converting a silicon strip waveguide mode give in Figure 2.3(a) in a gap SPP of a vertical MSW covered with an electro-optic cladding of M3. Figure 3.7 (b) gives the electric field distribution in such a mode converter. We design the MTMC for a silicon-on-insulator platform with a silicon device layer of 220 nm. We select a silicon strip waveguide width of 450 nm to have relatively small initial mode diameter. The MSW is made of gold electrodes with an optimum thickness of 150 nm as discussed in Section 2.2.3.

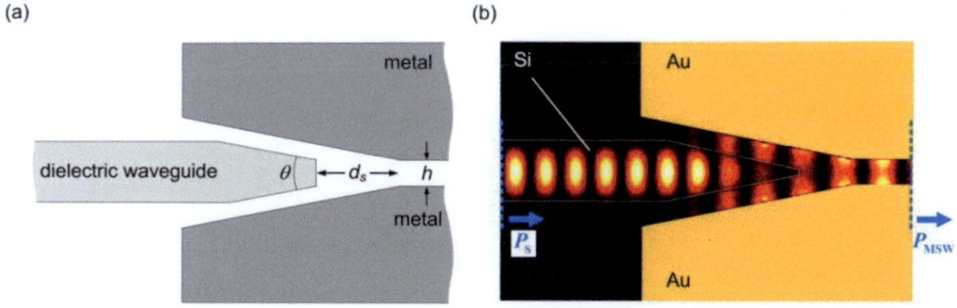

Figure 3.7 Metallic tapered mode converter (MTMC) that can be used to convert a dielectric waveguide mode in a gap SPP of a MSW.(a) Top view of a MTMC with a tapering angle of θ, metallic slot width of h. (b) Electric field distribution in a MTMC between silicon strip waveguide with a width of 450 nm and plasmonic metallic slot waveguides for a slot width of 140 nm.

The geometrical parameters that influence the photonic-to-plasmonic conversion efficiency are the tapering angle θ and the distance d_s shown in Figure 3.7 (a). We use a numerical solver employing a frequency domain finite integration technique to find the optimum values of θ and d_s which result in the maximum conversion efficiency [128], see also Appendix A.8. The power conversion efficiency is measured as the ratio of the input power from the silicon waveguide that is converted into the gap SPP,

$$\left(c_{\text{MWS}}^{\text{s}}\right)^2 = \frac{P_{\text{MSW}}}{P_{\text{s}}}. \tag{3.4}$$

More than 80 % of the input power can be converted in a gap SPP of MSWs with sub-200 nm slots. The results are shown in Figure 3.8. For a MSW with a slot width of 140 nm, we select an optimum distance d_s of 500 nm and tapering angle of 15°. This geometry shows relatively good tolerance with respect to the fabrication errors in defining the d_s and θ.

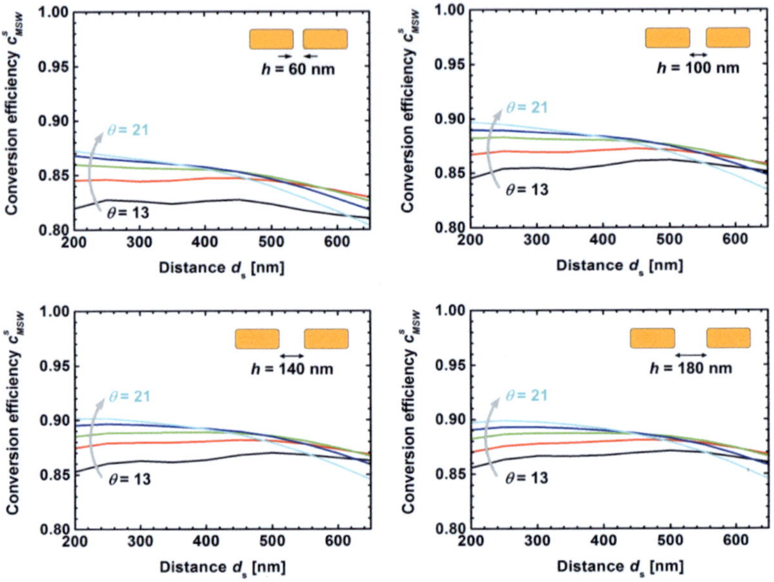

Figure 3.8 Amplitude conversion efficiency of a MTMC as a function of the distance d_s for five different taper angles $\theta = 13°$(black), 15°(red) up to 21°(cyan). The amplitude conversion efficiencies are given for vertical MWSs with a slot width h of 60 nm (a), 100 nm (b), 140 nm (c) and 180 nm (d).

3.2.2 Results and Discussion

We fabricated metallic tapered mode converters as described in Appendix B.1. Pairs of MTMCs are fabricated for both photonic-to-plasmonic and plasmonic-to-photonic mode conversions. By varying the length L_{MSW} of the MSW between the pairs of MTMCs we can extract the conversion efficiency similar to the standard cut-back measurement [129]. We successfully fabricated three devices with MSW lengths L_{MSW} of 1 µm, 29 µm, and 44 µm, see Figure 3.9. The slot width is about 140 nm for all three cases.

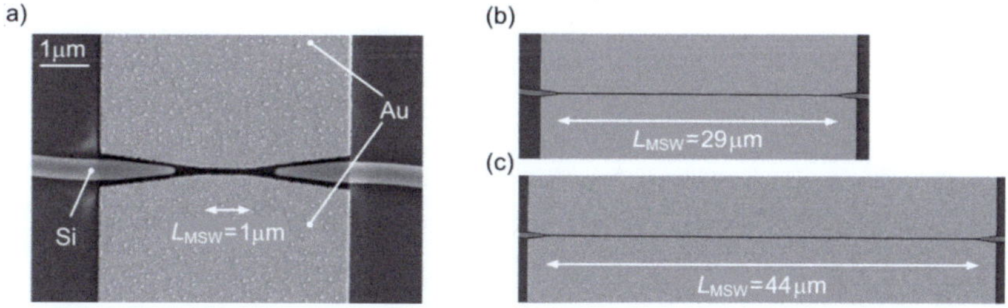

Figure 3.9 Fabricated metallic tapered mode converters with three different MSW lengths L_{MSW} of (a) 1 μm, (b) 29 μm, and (c) 44 μm. The slot size h is 140 nm for all three devices.

The measured silicon-to-silicon waveguide transmission spectra for the three different MSW lengths L_{MSW} are given in Figure 3.10. The measured transmission spectra are normalized to the measured reference spectra for a silicon strip waveguide without a plasmonic section. As can be seen, the tapered mode converters exhibit large conversion efficiency in a wide operating wavelength range. A total conversion loss of 1 dB is estimated for two transitions. This is in good agreement with the theoretically expected conversion efficiency of 2…3 dB. The difference between theoretically calculated and experimentally measured conversion efficiencies is attributed to small differences of the fabricated slot widths and variations of the sidewall roughness of the metallic slots.

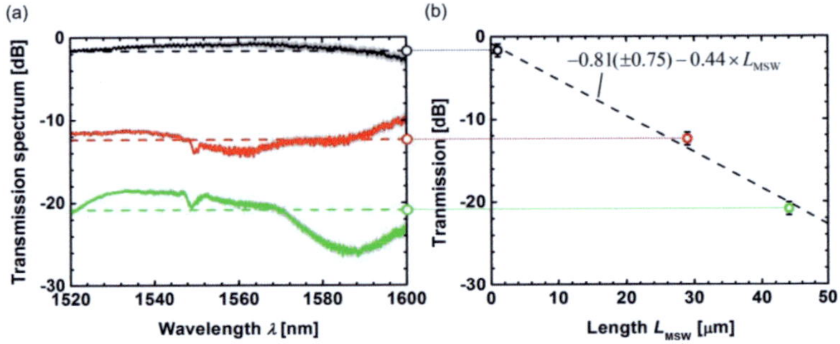

Figure 3.10 Measured transmission spectra of the fabricated MTMCs. (a) Transmission spectra for MTMCs with MSW lengths L_{MSW} of 1 μm (black), 29 μm (red) and 44 μm (green). Oscillatory features in the spectra are attributed to Fabry-Perot reflections in the device. (b) Power transmission versus device length L_{MSW}. The oscillations in the spectra have been averaged on a linear scale are taken for estimation of the coupling loss.

3.2.3 Conclusion

We optimized the MTMC design in order to convert the quasi-TE mode of a silicon strip waveguide in a gap SPP of a silicon MSWs filled with the electro-optic organic material M3 having a refractive index of 1.68. We fabricate and characterize MTMCs with MSWs with a slot width h of 140 nm. A mode conversion efficiency of about 1 dB is measured.

3.3 Plasmonic Polarization Beam Splitter

Parts of this section have been published in [C1].

A. Melikyan, C. Gärtner, K. Koehnle, A. Muslija, M. Kohl, C. Koos, W. Freude, J. Leuthold, "Integrated wire grid polarizer and plasmonic polarization beam splitter," in Optical Fiber Communication Conference (OFC'12), Los Angeles (CA), USA, paper OW1E.3, 2012.

[Start of Paper]

Abstract: Ultra-compact, low loss, high extinction ratio polarization beam splitter is proposed integrated on SOI platform. The device is 3.5 µm in length and provides more than 20 dB extinction ratio with plasmonic losses of 2 dB.

3.3.1 Introduction

Ultra-compact, lowest loss polarization beam splitters (PBSs) are vital elements that are needed if CMOS compatible silicon photonics is going to be deployed. The reason for the small footprint requirements on photonics are because of the expense associated with every square micrometer in a CMOS fab. Moreover, for the next generation of optical interconnects between chips employing polarization and wavelength diversity schemes, PBSs should provide broadband operation through the entire communication band. In addition, the PBSs should be simple to fabricate, and be compatible with silicon photonics.

Previously, polarization beam splitter cubes have been widely used for polarization multiplexing and demultiplexing due to their high extinction ratio (ER) and broadband operation [130]. Nowadays, polarization beam splitter cubes provide more than 20 dB extinction ratio in the entire communication band. Another historical approach to separate two orthogonal polarizations is the wire grid polarizer (WGP), first time reported for radio waves by Heinrich Hertz in 1888 , and afterwards applied to optical frequencies by C. H. Brown in 1940 [131]. Wire grid polarizers consist of metallic wires which are separated by a distance smaller than the wavelength of the incident electromagnetic (EM) wave. An EM wave polarized perpendicularly to the metallic wire grids passes through the polarizer, while the EM wave polarized parallel to the gird is partially reflected back and partially absorbed. However, these devices are bulky and not well suited for integrated silicon photonics.

More compact PBSs have been proposed making use of, e. g., photonic crystals [132–134], directional couplers [135], gratings [136, 137], and surface plasmon polaritons [138]. However, the above mentioned devices exceed the footprint of the PBS proposed in this work and require relatively complicated fabrication procedures. Recently, a polarization beam splitter consisting of a metal nano-ribbon and two dielectric waveguides was proposed and numerically investigated [139]. While the length of the PBS is shorter than 3 μm, the fabrication issues coming from the enormously large aspect ratio of the nano-ribbons are still to be solved.

Here, we propose a new, simple to fabricate, integrated plasmonic PBS with an ultra-compact size of 3.5 μm and with a high extinction ratio. Our PBS is compatible with the standard SOI platform. The scheme takes advantage of the fact that two orthogonal polarizations can efficiently be separated with negligible losses by making use of a vertical metallic slot waveguide (MSW). The operation principle is similar to the wire grid polarizers and provides broadband performance. Unlike to other published approaches, our PBS avoids unnecessary losses and resonances, and it has a very high directivity unlike polarization beam splitters based on photonic crystals [132–138].

3.3.2 Design and Analysis

The polarization beam splitter is built on a silicon dioxide substrate, see Figure 3.11. It consists of two parallel silicon strip waveguides (SWG) separated by 200-300 nm. Each SWG has a thickness of 340 nm and a width of 300 nm. The cladding material is taken to be silicon dioxide with a refractive index of 1.44. Similar performance can be also achieved in the case of a PMMA cladding, because PMMA and silicon dioxide have similar refractive indices. A vertical MSW consisting of two silver bars is placed between silicon waveguides. The MSW supports a highly confined gap SPP which has a dominant electric field along the y-axis. This gap SPP, therefore, can be excited only via a quasi-TM mode of the SWG as discussed in Section 3.1. A quasi-TE mode of the SWG can excite SPPs at the sidewalls of the metal bars only (i. e., at the SiO_2-silver interface). This SPP does not significantly extend into the slot. Thus, in our case, the MSW can act as a polarization selective channel between two SWGs.

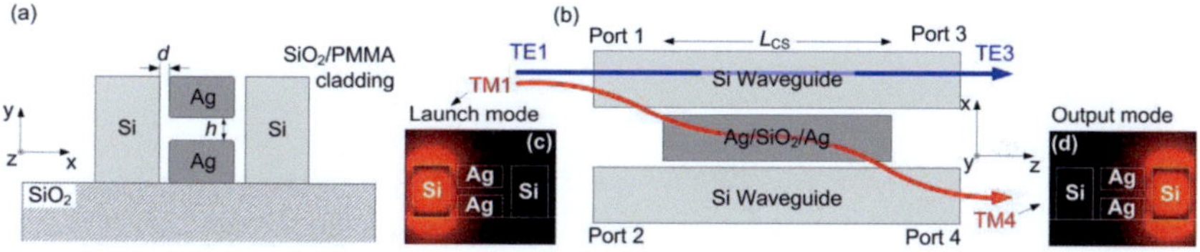

Figure 3.11 Schematic of the polarization beam splitter. (a) Cross-section and (b) top view of the PBS. When launching a quasi-TM mode (c) from Port 1, light couples to the right SWG through the MSW and propagates to Port 4. The quasi-TE mode launched from Port 1 passes through the device almost undisturbed and couples to Port 3.

When a quasi-TE mode enters at Port 1, the two vertically stacked metal bars act as one, because a SPP cannot be excited, and they prevent the incident wave from coupling to the neighboring SWG. Therefore, the quasi-TE mode can pass undisturbed to Port 3. For a quasi-TM excitation, however, the wave from the left SWG excites an SPP which couples to the right SWG. Thus, the MSW connects the two quasi-TM modes of the SWGs. Choosing an appropriate cladding and the proper waveguide dimensions, the coupling efficiency of two quasi-TM modes of the SWGs can significantly be increased, leading to a high polarization extinction ratio.

The PBS length and the coupling efficiency of the quasi-TM mode at Port 1 to the quasi-TM mode at Port 4 are calculated using the simplified eigenmode expansion method (EME) described in Section 3.1.2. First, we calculate the quasi-TM mode at Port 1, see Figure 3.11(c). Second, the supermodes of the coupling section are simulated. Using Eq. (3.1) and Eq. (3.2) the overlap integral $c_\mu^{\mathrm{TM}1}$ between the quasi-TM mode of Port 1 and each supermode μ is calculated. Six supermodes of the coupling section are given in Figure 3.12 with their respective excitation efficiencies $c_\mu^{\mathrm{TM}1}$.

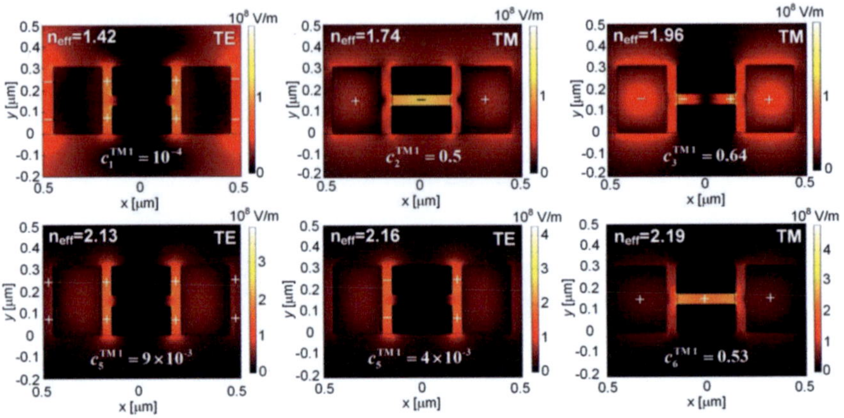

Figure 3.12 Supermodes of the polarization beam splitter (PBS) with $d = 50$ nm and $h = 50$ nm.The amplitude coupling efficiency c_{μ}^{TM1} for quasi-TM excitation at Port 1 is given below each mode profile. The relative phase of the dominating electric field component in the cross-section (E_y or E_x for quasi-TE or quasi-TM, respectively) is represented with (+) and (−)-signs. Second, third and sixth supermode are efficiently excited by a quasi-TM at port 1 (left-hand side SWG).

The field distribution in the PBS due to the propagation of these supermodes can be expressed as in Eq. (3.3) by replacing s with TM 1,

$$\begin{aligned} \breve{\mathbf{E}}_{\mathrm{CS}}(\mathbf{r},\omega) &= \sum_{\mu=1}^{N} c_{\mu}^{\mathrm{TM1}} \breve{\mathbf{e}}_{\mu}(x,y,\omega)\cdot e^{-\mathrm{j}\beta_{\mu}z}, \\ \breve{\mathbf{H}}_{\mathrm{CS}}(\mathbf{r},\omega) &= \sum_{\mu=1}^{N} c_{\mu}^{\mathrm{TM1}} \breve{\mathbf{h}}_{\mu}(x,y,\omega)\cdot e^{-\mathrm{j}\beta_{\mu}z}. \end{aligned} \tag{3.5}$$

The excitation efficiency of μth supermode is c_{μ}^{TM1} and $\breve{\mathbf{e}}_{\mu}$ and $\breve{\mathbf{h}}_{\mu}$ are the complex electric and magnetic fields of the μth supermode, respectively. In our case, the summation includes $N = 10$ supermodes.

Third, the normalized overlap integral $c_{\mathrm{TM4}}^{\mathrm{CS}}(z)$ between the field distribution Eq. (3.5) and the quasi-TM mode of the right SWG is calculated as a function of the z-coordinate using again Eq. (3.2) by replacing j with CS and i with TM 4. The resulting amplitude coupling coefficient $c_{\mathrm{TM}\,4}^{\mathrm{CS}}(z)$ represents the coupling efficiency of the quasi-TM modes at Port 1 and at Port 4 as a function of length $z = L_{\mathrm{CS}}$ of the straight coupling section, excluding the waveguide bends. The larger $c_{\mathrm{TM}\,4}^{\mathrm{CS}}(L_{\mathrm{CS}})$, the larger the coupling coefficient $c_{\mathrm{TM}\,4}^{\mathrm{TM}\,1}$ of the quasi-TM mode at Port 1 to the quasi-TM mode at Port 4 would be, and the smaller the coupling is to Port 3, meaning that the polarization extinction ratio becomes high. To achieve this, the coupling length L_{CS} should be optimized.

3.3.3 Results and Discussion

We plot $c_{\mathrm{TM}\,4}^{\mathrm{TM}\,1}$ and L_{CS} as a function of the distance d for three different slot sizes h of 30 nm (black), 40 nm (red), and 50 nm (green), see Figure 3.13. The quasi-TM mode at Port 1 can almost completely be coupled to Port 4 with a device length in a sub-10 μm length, see Figure 3.13.

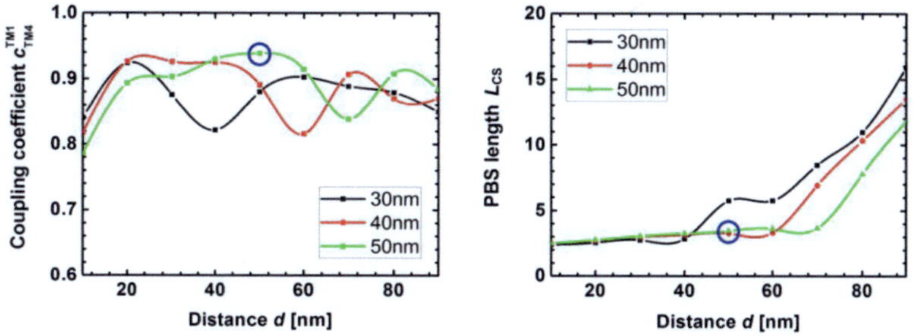

Figure 3.13 (a) Coupling efficiency of the quasi-TM mode at Port 1 to the quasi-TM mode at Port 4 and (b) the corresponding coupling length L_{CS}. The quasi-TM mode at Port 1 can be efficiently coupled to Port 4 in a PBS with a sub-5 μm length. Geometrical properties taken within the blue circles are considered in the operating bandwidth study by performing finite integral technique simulations.

To check the optical bandwidth of the designed PBS, we carried out full wave finite-integration simulations for the structure depicted in Figure 3.12. Figure 3.14(a) and Figure 3.14(b) gives the magnetic field distribution in the PBS for both quasi-TE and quasi-TM excitation. It can be seen that the quasi-TM mode excited from the left SWG couples through the MSW and reaches Port 4, while the quasi-TE mode stays in the left SWG. Figure 3.14(c) and Figure 3.14(d) show that at a wavelength of 1550 nm, a polarization splitting with an extinction ratio of about 20 dB is possible for a device length of less than 5 μm. Moreover, our PBS exhibits a broad optical operating range of at least 40 nm, and this is actually required in high-capacity short range as well as in long-haul telecommunications links when both polarization and wavelength diversity schemes are employed.

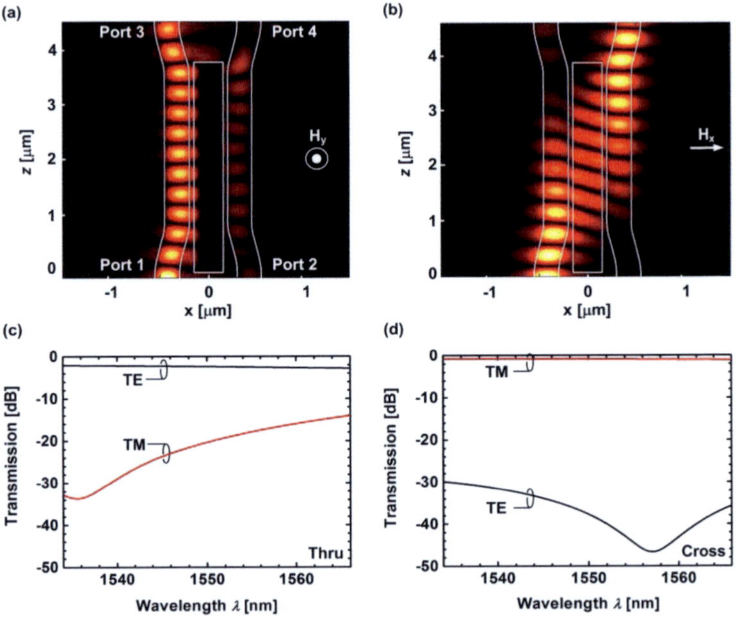

Figure 3.14 Field distribution in the polarization beam splitter (PBS) and transmission spectra at Port 3 and Port 4 for both TM and TE excitation. Magnetic field distribution in the PBS for the case of TE (a) and TM (b) excitations at a wavelength of 1550 nm. The quasi-TM mode launched from Port 1 couples through the MSW reaches Port 4, while the quasi-TE mode passes through the PBS, but remains in the left-hand side SWG. Transmission spectra at Port 3 (c) and Port 4 (d). In a PBS with a coupling length of only 3.5 μm, exinction ratios of more than 20 dB are feasiblity across a broad wavelength range.

3.3.4 Conclusion

We propose an ultra-compact, low loss, broadband and high extinction ratio plasmonic polarization beam splitter integratable on the SOI platform. The device requires simple fabrication approach and is well suited for on-chip polarization-multiplexed WDM systems.

[End of Paper]

4 Active Plasmonic Components

In this chapter, we firstly present the concept of plasmonic phase modulator based on the Pockels effect of electro-optic organic material. We discuss various characteristic of the proposed modulator and demonstrate its operation at the data rate of 40 Gbit/s. In the last part of this chapter a surface plasmon polariton absorption modulator is presented and theoretically studied. Characterization results of a surface plasmon polariton absorption modulator fabricated on a silicon-on-insulator platform are reported.

4.1 SPP Phase Modulator

The content of this section is a direct copy of the Journal publication [J2].

A. Melikyan, L. Alloatti, A. Muslija, D. Hillerkuss, P. C. Schindler, J. Li, R. Palmer, D. Korn, S. Muehlbrandt, D. Van Thourhout, B. Chen, R. Dinu, M. Sommer, C. Koos, M. Kohl, W. Freude and J. Leuthold, "High-speed plasmonic phase modulators," Nat. Photonics 8, 229–233, (2014).

Minor changes have been done to adjust position of figures and notations of variables.

[Start of Paper]

To keep pace with the demands in optical communications, electro-optic modulators should feature large bandwidths, operate across all telecommunication windows, offer a small footprint, and allow for CMOS compatible fabrication to keep cost low [11]. And while many platforms have been tried, plasmonics has been forecast as a potential solution. Here we demonstrate a new ultra-compact plasmonic phase modulator based on the Pockels effect in a nonlinear organic material. The device has a length of only 29 μm and operates at 40 Gbit/s. Its modulation frequency response is flat up to 65 GHz and beyond. The modulator has been tested to work across a 120 nm wide wavelength range centred at 1550 nm, and is expected to work also way beyond this range. Its operation has been verified for temperatures up to 85° C, and it is easy to fabricate. To the best of our knowledge, this is the most compact high-speed phase modulator demonstrated so far.

4.1.1 Introduction

High speed, ultra-compact and power efficient electro-optic modulators are currently in the spotlight of research as they are key components in optical transmission link [11, 77]. To this day the majority of silicon-based modulators exploit either the plasma dispersion effect in silicon [27, 30, 31, 33, 34, 40, 41, 44, 45, 140] or the Pockels effect in a nonlinear cladding, a technology also known as silicon-organic hybrid (SOH) [18, 19, 21]. The configuration of silicon modulators can be classified as resonant [19, 27, 30, 31, 40, 41, 44] or non-resonant [18, 33, 34, 45, 140]. Resonant modulators can be very compact. A device with a footprint as small as 78 μm^2 has already been shown [27]. This size is due to the large quality factor of the resonant cavity, which enhances nonlinear interaction by several orders of magnitude. However, resonant modulators suffer from bandwidth limitations and need to be optimized for a certain operating wavelength range. In addition, resonant modulators are sensitive to temperature fluctuations and fabrication tolerances. In contrast, non-resonant modulators offer operation across a large spectral window. They are typically based on a traveling wave configuration. To achieve a long interaction time between the optical and the modulating RF signal these devices are often millimetres in length which increases the RF losses. Such bulky dimensions prevent also an economically efficient co-integration with electronics. A further disadvantage of the travelling wave approach is the requirement of terminating the RF transmission line with matched impedance, which reduces the modulating voltage to half of the generator's open circuit voltage.

Surface plasmon photonics or "plasmonics" is the art of controlling, guiding and detecting surface plasmon polaritons (SPPs). The SPP is an electromagnetic surface wave at a dielectric-metal interface, coupled to the charge density oscillation in the metal surface [1, 2]. Plasmonics is a means of providing ultra-compact and high-speed components such as modulators, switches and detectors [116, 141–143]. Since the 1980s the realization of a plasmonic modulator was in the focus of many research groups [16, 24, 25, 37, 54, 68, 120, 123]. Modulation of an SPP with an electrical signal has been shown by employing the thermo-optic effect [16], the Pockels effect [24, 25, 120], and the plasma dispersion effect either in semiconductors or in metals [37, 54, 68, 123]. However, despite the many approaches pursued in the past 30 years, an electro-optical high-speed plasmonic modulator has never been shown.

Here, we present a plasmonic phase modulator (PPM) that combines the large RF bandwidth of a traveling wave modulator [18, 21, 33, 34, 45, 140], the compactness of a photonic crystal modulator [19, 44] and the energy efficiency of a ring modulator [27, 30, 31, 40, 41]. In particular, the reported PPM has an RF bandwidth, which exceeds the bandwidth of known silicon modulators [18, 19, 21, 27, 30, 31, 33, 34, 40, 41, 44, 45, 140], while its footprint is still among the smallest [19, 27, 30, 31, 40, 41, 44].

4.1.2 Design and Fabrication

Our PPM consists of two metal tapers performing the photonic-to-plasmonic mode conversion, and a phase modulator section in-between, see Figure 4.1(a). The metal taper narrows down at an angle of 15° and is used as an interface between silicon photonics and plasmonics [101]. Light guided through the silicon nanowire efficiently excites the SPP via the metal taper. The SPP is then guided into the phase modulator section, which consists of two metal pads separated horizontally by a nanometer-scale vertical slot, see Figure 4.1(b), (c). The slot is filled with a nonlinear organic material, the refractive index n of which can be changed via the Pockels effect by applying a static electric field U/h (modulating voltage U, gap width h). With the electro-optic coefficient r_{33}, the change of the refractive index amounts to $\Delta n_{\mathrm{EO}} = (1/2) \times r_{33} n_{\mathrm{EO}}^{3} U/h$ [82]. By modulating the refractive index of the polymer in the slot, the information is encoded in the phase of the SPP. At the end of the modulator section, the SPP is back-converted into a photonic mode.

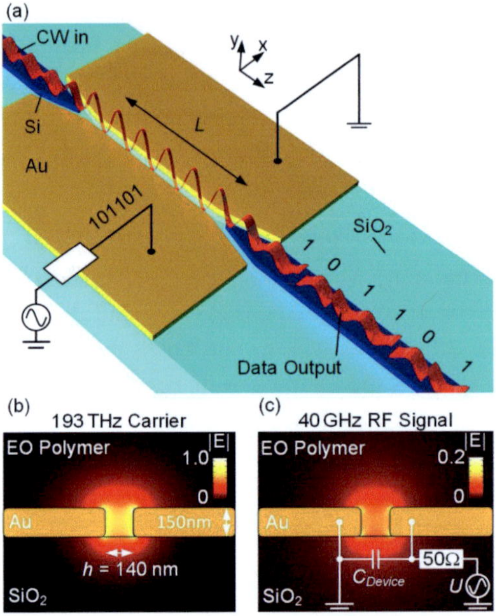

Figure 4.1 Plasmonic phase modulator and field distributions. (a) Schematic of plasmonic phase modulator. Continuous-wave infrared light guided by the upper-left silicon nanowire is coupled through a metal taper to the plasmonic slot waveguide. The slot in the metal sheets is filled with a linear electro-optic (Pockels effect) nonlinear organic material. The phase of the surface plasmon polariton (SPP), which propagates in the slot, is changed by applying a modulating voltage. A second taper transforms the SPP back to a photonic mode in the lower-right nanowire. Mode profiles of the SPP (b) and of the RF-signal (c), respectively, showing the colour-coded modulus of the complex electric field vector.

Noble metals such as gold (Au) or silver (Ag) exhibit negative dielectric permittivity below their plasma frequency. For such metals, the modulus of the permittivity at communication wavelengths is typically two orders of magnitude larger than the permittivity of the nonlinear organic material [86]. This large permittivity contrast at the metal-polymer interface results in an optical field enhancement in the slot, see Figure 4.1(b). Moreover, both optical and modulating fields are strongly confined to the slot, resulting in a near-perfect overlap between optical and RF signal, see Figure 4.1(b) and Figure 4.1(c). Consequently, the U_π

voltage for a phase shift of π at a given device length L is small, leading to a small voltage-length product $U_\pi \times L$. Therefore, a sub-50 µm PPM with a sufficient modulation index η (unit rad) becomes feasible. In our device, we employ the Pockels effect in a nonlinear organic material, thereby avoiding the speed limitation typically associated with the carrier lifetime in plasma-effect based injection-type modulators [27, 34, 40]. High conductivity of metal sheets and the small capacitance of the device result in an ultra-small RC time constant which does not pose a practical speed limitation [18, 19, 21, 27, 30, 31, 33, 34, 40, 41, 44, 45, 140]. Furthermore, the lumped structure of our PPM avoids bandwidth limitation caused by the walk-off of electrical and optical signals. The device's purely capacitive nature and the small U_π make it also energy efficient. Optical broadband operation of the PPM is ensured by the weak wavelength-dependence of the electro-optic organic material, by the inherently broadband metallic slot waveguide, and by the metallic tapers.

4.1.3 Characterization

We fabricate two generations of PPM on the SOI platform, see Appendix B.1. The first generation (1stG) of PPM has been fabricated in a ground-signal-ground (GSG) configuration with metallic slots, having a length of 34 µm and a slot width of 200 nm, on each side of the signal electrode. In the second generation (2ndG) PPM, the slot size has been reduced down to 140 nm, and the length was reduced to 29 µm. The 2ndG device was designed to have a ground-signal (GS) configuration. A scanning electron microscope (SEM) picture of the 1stG device is shown in Figure 4.2(a).

We measured the optical loss of the modulator section by taking an equal-length SOI strip waveguide as a reference. The fabricated PPMs show broadband optical transmission, Figure 4.2(b). The average total loss is 12 dB for both modulators. The electrical properties of the 1stG modulator are studied by investigating the S_{11} RF reflection factor. The cross-section of the device and its equivalent electrical circuit are given in Figure 4.2(c). An RF reflection factor $|S_{11}|$ is measured to be larger than 95 % across a 60 GHz frequency range, see Figure 4.2(d). We fit the measured electrical reflection factor with a lumped equivalent circuit depicted in Figure 4.2(c) and find that the capacitance of each of the metallic slots is $C_{1stG} \approx 2.2$ fF. By geometrically scaling with respect to the slot width (140 nm vs. 200 nm) and to the length (29 µm vs. 34 µm) we estimate the capacitance of the 2ndG to be $C_{2ndG} \approx 2.7$ fF.

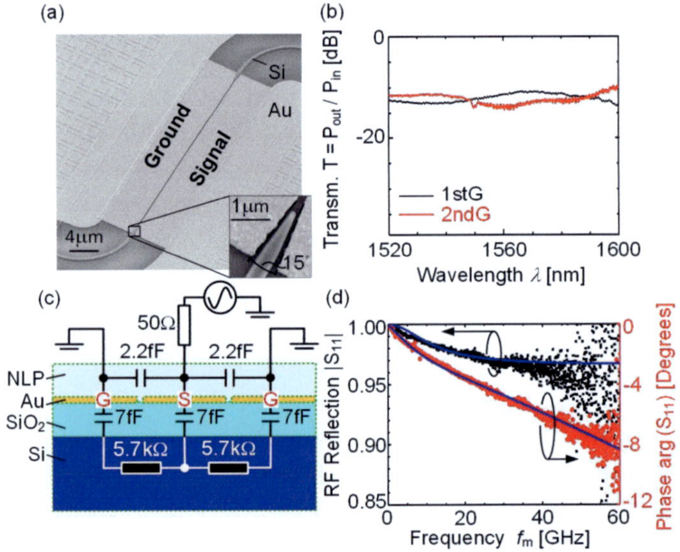

Figure 4.2 Characteristics of fabricated plasmonic phase modulators. (a) Scanning electron microscope (SEM) picture of the 1stG plasmonic phase modulator with a length of 34 µm. The photograph is taken before coating the sample with an electro-optic polymer. The lower-right inset shows the clearly defined silicon nanowire taper as well as the good alignment of taper and slot. (b) Comparison of the measured optical transmission of the 1stG device (black solid curve) with 2ndG modulators (red solid line) after coated with an electro-optic polymer. A flat transmission across a wide spectral range is seen. (c) Lumped element model of the 1stG modulator used for electrical characterization with a ground-signal-ground (GSG) probe. (d) Magnitude and phase of the S_{11} RF reflection factor for frequencies up to 60 GHz. In addition, the modulus (left axis) and the phase of S_{11} (right axis) as calculated for the lumped element model depicted in **c** is given (blue solid line).

We characterized the electro-optic frequency response of the modulators using the measurement setup given in the inset of Figure 4.3(a). The modulators are driven with a sinusoidal RF signal with an amplitude of U_m and the resulting modulation index is measured, see Appendix B.2. The measured phase modulation index for the 2ndG modulator is shown in Figure 4.3(a) as a function of the RF modulation frequency up to 65 GHz. To cover the entire RF frequency range of the driver we have selected relatively small drive voltage amplitude of $U_m = 0.1V$. The device exhibits ultra-flat frequency response up to at least 65 GHz. Such flat response was only possible due to the instantaneous Pockels effect and the ultra-small *RC* time constant of the device. For characterizing the optical response we measure the phase modulation index for various carrier wavelengths λ_c keeping the modulation frequency at 45 GHz and the drive voltage amplitude $U_m = 0.8V$ constant. Figure 4.3(b) shows the results for λ_c between 1480 nm and 1600 nm. Fitting the measured modulation index η with the one predicted with the first order perturbation theory (see Appendix A.7) we derive the achieved on-chip electro-optic coefficient of the polymer. The electro-optic coefficients obtained for the 1stG and 2ndG devices are $r_{33} = 13$ pm/V and $r_{33} = 21$pm/V, respectively.

Even though the used nonlinear organic material has been previously tested in a high temperature environment [144], we carried out thermal stability test to further confirm the device's operation under operation conditions such as found e.g. in telecommunications racks. The modulation index has been measured for various sample temperatures up to 85° C. No degradation of the modulation index has been found, see Figure 4.3(c).

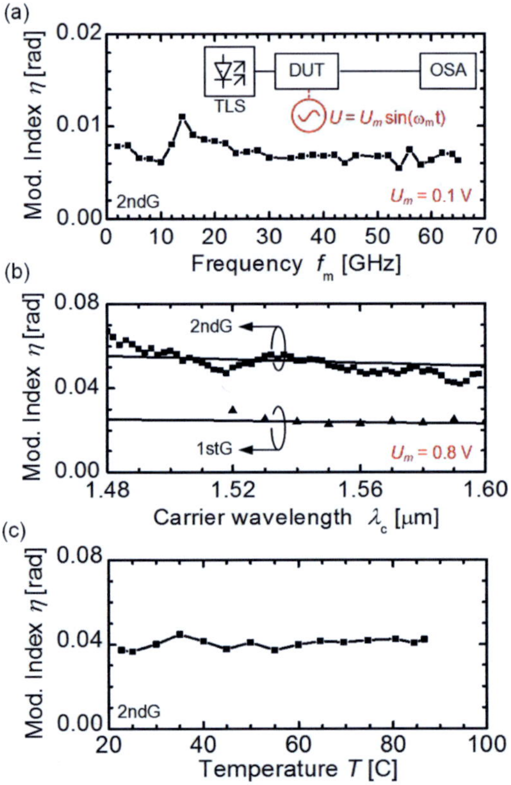

Figure 4.3 Measured performance of the plasmonic phase modulators. (a) Modulation index η versus RF frequency f_m for 2ndG modulator. The frequency response is flat up to a frequency of at least 65 GHz. Experimental setup used for measuring the phase modulation index η for various RF and carrier frequencies is given as an inset. Infrared light from the tuneable laser source (TLS) is sent to the device under test (DUT). The modulator is driven by a sinusoidal electrical signal with an amplitude of U_m and the optical spectrum is measured at the output of the chip with an optical spectrum analyzer (OSA). (b) Modulation index for 1stG and 2ndG modulator versus optical carrier wavelength in a range of 120 nm for a modulation frequency of $f_m = 45$ GHz. The modulation index is essentially flat with respect to the carrier wavelength in a range of 120 nm. Solid lines give the modulation index predicted by first order perturbation theory. (c) Thermal stability of the device is tested for temperatures up to 85 C. The polymer exhibits no degradation of its electro-optic coefficient in high temperatures similar what has been shown elsewhere [144].

4.1.4 High-speed Data Modulation

To test the high-speed capabilities of the 1stG modulator we encode a data stream with a bitrate of up to 40 Gbit/s, the maximum rate available in our lab. The phase of the SPP is encoded with a $2^{31} - 1$ long pseudo-random bit sequence (PRBS) at a voltage swing of $U_{pp} = 7.5$ V, resulting in a peak-to-peak phase modulation of 0.23 rad. The resulting binary phase shift keyed (BPSK) signal is received by a coherent receiver. The error vector magnitudes (EVM) for various data rates are given in Figure 4.4(b), (d). The bit error ratios (BER) corresponding to the measured EVMs [145] are in the range of $(3\ldots7) \times 10^{-3}$.

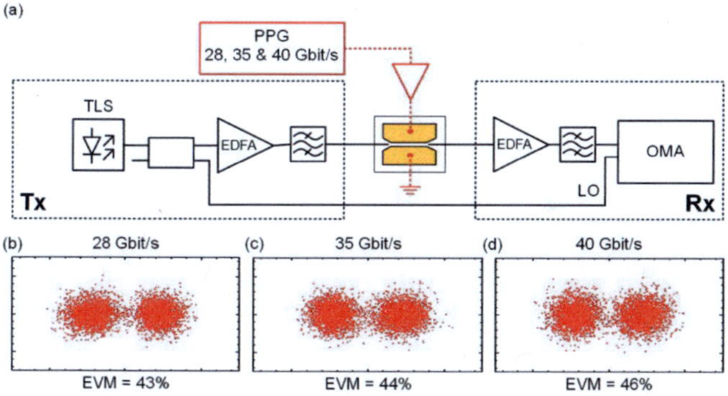

Figure 4.4 Modulation experiments with the 1stG plasmonic phase modulator. (a) Experimental setup for measuring the bit error ratio (BER) and the error vector magnitude (EVM) of a binary phase shift keyed (BPSK) signal. Light with a wavelength of 1550.92 nm is amplified by an erbium-doped fibre amplifier (EDFA), filtered, and sent to the DUT, wherein the phase of the SPP is modulated by the electrical RF signal. The resulting optical BPSK signal is detected with an Agilent Optical Modulation Analyser (OMA). (b), (c), (d) Constellation diagrams of the received BPSK signal for the data rates 28 Gbit/s, 35 Gbit/s, and 40 Gbit/s. EVM between 43 % and 46 % correspond to BER of $(3\ldots7) \times 10^{-3}$.

The 2ndG modulator is encoded with a 40 Gbit/s data stream. In this case, the device has been operated with an electrical signal with a reduced amplitude of $U_{pp} = 4.7$ V, resulting in a peak-to-peak phase modulation of 0.31 rad. This signal was subsequently converted into an intensity modulated signal by means of a delay interferometer (DI) with a free spectral range (FSR) of 40 GHz, see

Figure 4.5(a). The signal is then directly detected with a single photodiode. A measured eye diagram with a BER of 6×10^{-5} is given in Figure 4.5(b). An optical physical link with such a BER can be used in both long and short range applications [146]. Taking into account the capacitance C_{2ndG} of the modulator we estimate an energy consumption per bit of $(2U_{pp})^2 \times C_{2ndG} / 4 = 60$ fJ/bit for a voltage swing $2U_{pp}$ across the device [147].

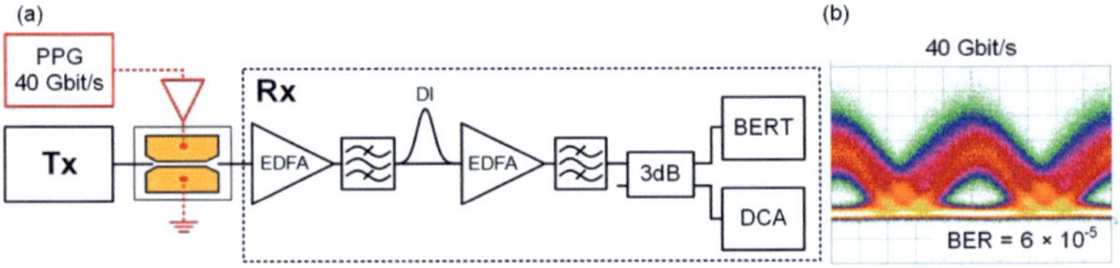

Figure 4.5 Modulation experiments with the 2ndG plasmonic phase modulator. (a) Direct receiver setup for detecting a binary phase shift keyed (BPSK) signal. 40 Gbit/s BPSK signal is converted in an intensity modulated signal using a delay interferometer (DI) with a free spectral range of 40 GHz. To compensate the losses of the in-house built DI, the second preamplifier is added Digital communication analyzer (DCA) and bit error ratio tester (BERT) are taken for eye diagram and BER measurement, respectively. (b) Eye diagram of the received signal for the data rate of 40 Gbit/s with a corresponding BER of 6×10^{-5}. Employing the latest electro-optic organic material [76] the average energy per bit could be reduced down to 18 fJ/bit for a plasmonic Mach-Zehnder Modulator (MZM) operating in push-pull, as we discuss in Chapter 1.

4.1.5 Conclusion

In conclusion, we report on the first successful experimental demonstration of a high-speed plasmonic phase modulator. We demonstrate operation at a high bit-rate of 40 Gbit/s for a device length of only 29 µm. The modulator exhibits a flat modulation frequency response for at least up to 65 GHz. It can be operated across the whole optical S, C and L-bands. Thermal stability of the device for temperatures up to 85°C has been assured. We believe that the concept has the potential to pave the way to future ultra-compact and CMOS compatible on-chip MZ and IQ plasmonic modulators such as needed to generate quadrature amplitude modulation formats.

[End of Paper]

4.2 POH modulators for OOK and BPSK signaling at 40 Gbit/s

The content of this section is a direct copy of the Journal publication [J3].

A. Melikyan, K. Koehnle, M. Lauermann, R. Palmer, S. Koeber, S. Muehlbrandt, P. C. Schindler, D. L. Elder, S. Wolf, W. Heni, C. Haffner, Y. Fedoryshyn, D. Hillerkuss, M. Sommer, L. R. Dalton, D. Van Thourhout, W. Freude, M. Kohl, J. Leuthold and C. Koos, "High-speed plasmonic phase modulators," Opt. Express, vol. 23, no 8, pp. 9938–9946, (2015)

Minor changes have been done to adjust position of figures and notations of variables.

[Start of Paper]

Abstract: We report on high-speed plasmonic-organic hybrid Mach-Zehnder modulators comprising ultra-compact phase shifters with lengths as small as 19 µm. Choosing an optimum phase shifter length of 29 µm, we demonstrate 40 Gbit/s on-off keying (OOK) modulation with direct detection and a BER < 6×10^{-4}. Furthermore, we report on a 29 µm long binary-phase shift keying (BPSK) modulator and show that it operates error-free (BER < 1×10^{-10}) at data rates up to 40 Gbit/s and with an energy consumption of 70 fJ/bit.

4.2.1 Introduction

Short and mid-range optical links utilizing wavelength division multiplexed (WDM) demand high-speed and compact integrated light modulators, which operate in a broad optical wavelength range at an energy consumption of a few femtojoules per bit [148]. Compact integrated modulators are commonly realized on the silicon photonic platform using resonant structures such as ring or disk resonators, or as photonic crystals [11, 14, 19, 30, 33, 35, 44, 77, 149, 150]. These devices exploit resonant enhancement of the light-matter interaction to enable compact implementations and low driving voltages. As an example, a modulator with a footprint of only 18…25 µm^2 has been reported employing microdisk resonators [149]. However, ring or disk modulators require careful matching of the carrier wavelength and of the resonance wavelength, by, e.g., heaters, which comes along with additional power consumption. Moreover, these devices do not open an easy route to advanced modulation formats such as quadrature amplitude modulation (QAM) [151]. As an alternative, photonic crystal (PC) modulators [19, 44, 150] can be used, allowing lengths of down to 50 µm [44]. However, PC based modulators usually occupy larger chip areas than ring-resonator modulators due to the large transverse dimensions of the PC sections. Furthermore, both for ring resonators and photonic crystal modulators, the optical bandwidth is limited by their resonant nature. These limitations can be overcome by plasmonic integration, which enables compact modulators without making the devices resonant [24, 37, 55, 68, 118, 122, 123, 152].

A high-speed plasmonic phase modulator with a length of 29 µm and with large RF and optical bandwidths have recently been reported by our group, exploiting the benefits of plasmonic-organic hybrid (POH) integration [122]. This approach is based on combining plasmonic slot waveguide with electro-optic (EO) organic cladding materials, thereby enabling binary phase-shift keying (BPSK) at 40 Gbit/s.

However, the application of plasmonic modulators within short-reach optical links calls for much lower BERs in order to minimize coding overhead for forward error correction (FEC), high-speed signal processing and the associated power consumption as well as associated latency. Furthermore, on-off-keying (OOK) with direct detection remains the most promising method for short-reach transmissions because of its technical simplicity and low implementation costs [153]. Thus, a plasmonic MZM that can reliably generate on-off keying (OOK) signals is in the focus of interest.

Here, we report on highly efficient POH modulators that enable the lowest BER that has so far been achieved with plasmonic devices at technically relevant data rates. In particular, applying a cladding of state-of-the-art organic material YLD124 / PSLD41 on our 29 µm-long phase modulator leads to a voltage-length product $V_\pi \times L$ of 0.45 V mm. Using this device, we demonstrate error-free (BER $< 1\times10^{-10}$) binary-phase shift keying (BPSK) signaling at data rates up to 40 Gbit/s. This BER is 5 orders of magnitude smaller than shown by us previously [122]. No FEC is required, and the modulator's energy consumption amounts to 70 fJ/bit. Furthermore, we report on the generation and direct detection of 40 Gbit/s OOK signals using plasmonic-organic hybrid (POH) Mach-Zehnder modulators (MZM). We test different devices featuring footprints of $21\times19\ \mu m^2$, $21\times29\ \mu m^2$ and $21\times39\ \mu m^2$. We discuss the optimum length of a POH MZM and find that 29 µm long plasmonic phase shifter (PS) sections represents an optimum in the sense that the necessary extinction ratio is achieved at an acceptable modulator insertion loss. This device allows generation of 30 Gbit/s, 35 Gbit/s and 40 Gbit/s OOK signals with a BER $< 6\times10^{-4}$ that is below the threshold of hard-decision FEC codes. Disregarding FEC, the energy consumption of these POH MZMs is in the range of 75 … 225 fJ/bit, depending on the device capacitance which is determined by the length of the PS. To the best of our knowledge, our experiments demonstrate the first error-free generation of a BPSK signal using a plasmonic modulator, and the first example of 40 Gbit/s OOK signaling using a plasmonic MZM.

4.2.2 Design and Fabrication

We realized our plasmonic modulators on the silicon photonic platform. The devices are based on a silicon-on-insulator wafer having a 2 µm thick buried oxide and a silicon device layer with a thickness of 220 nm. Silicon nanowire waveguides with a width of 450 nm and a height of 220 nm are used as access waveguides to the phase modulator sections. Light is coupled in and out of the silicon nanowires using silicon grating couplers. The BPSK modulator is constructed of a single high-speed plasmonic phase shifter (PS) operating as a phase modulator (PM), see Figure 4.6(a) and Figure 4.6(b) [122]. Figure 4.6(a) gives an artist's impression of the POH PM. Mode conversion between the quasi–TE mode of the silicon waveguide and the gap surface plasmon polariton (SPP) is accomplished via a tapered silicon waveguide enclosed by a tapered gap plasmon waveguide [101]. The phase of the optical signal is modulated in the 29 µm long plasmonic PM section, labelled as "Phase Modulator (PM)" in Figure 4.6(a) and "PM" in Figure 4.6(b). The POH MZM for OOK consists of one PS placed in each of the two arms of a Mach-Zehnder interferometer (MZI), which is realized on a silicon planar lightwave circuit (PLC), see Figure 4.6(c) and Figure 4.6(d). The MZI is designed with unbalanced arms to enable adjustment of the operating point by wavelength tuning. Standard silicon multimode interference (MMI) coupler with insertion loss of 0.2 dB serves as optical power splitter / combiner. High-speed phase modulation is performed by the PS exploiting the Pockels effect in an electro-optic organic cladding material [122]. Applying a voltage between the metal electrodes changes the refractive index of the electro-optic (EO) material in the slot due to the Pockels effect, and therefore influences the phase velocity of the plasmonic mode.

Both, POH-PM and MZM, are fabricated on the same SOI wafer. First, the silicon PLC is fabricated using standard processes such as 193 nm DUV lithography and Si dry etching in the framework of ePIXfab. The metallic slots having widths of ~150 nm are defined on the silicon PLC with electron-beam lithography and a lift-off process with poly-methyl-methacrylate (PMMA) resist. The PM is fabricated with a ground-signal configuration, see Figure 4.6(a) and Figure 4.6(b) [122]. For the MZM, two high-speed PS with a common signal electrode are fabricated in ground-signal-ground configuration as depicted in Figure 4.6(c) and Figure 4.6(d). The devices are fabricated on gold because of the low plasmonic losses and the expected long term stability. In the case of the POH PMs, the slot is filled with an electro-optic (EO) organic material YLD124/PSLD41(25:75 wt.%) mixture which has been previously shown to provide a record EO coefficient of $r_{33} = 230$ pm/V [154]. In the case of the POH MZM, we employ the EO material SEO100 (Soluxra, LLC) because of its excellent thermal stability up to 85°C, which is an important requirement for use in real-word communication systems. The electro-optic effect in the EO materials is activated through a poling procedure [154].

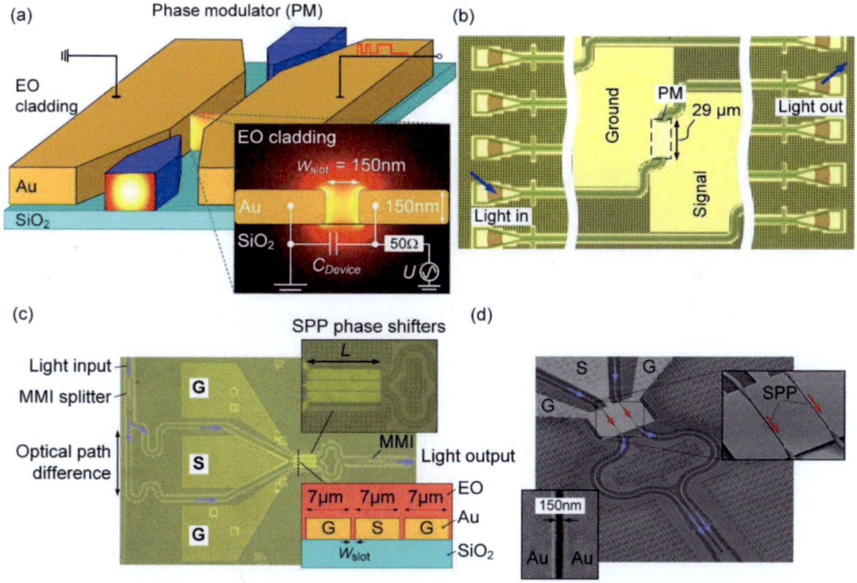

Figure 4.6 Plasmonic-organic hybrid (POH) phase modulator (PM) and Mach-Zehnder modulator (MZM), fabricated on the silicon-on-insulator (SOI) platform. (a) Schematic of the POH PM comprising a metallic slot waveguide filled and clad with an electro-optic (EO) material. The photonic mode of a silicon nanowire waveguide (blue) is converted to a gap surface plasmon polariton (SPP) via a tapered silicon waveguide enclosed by a tapered gap plasmon waveguide [101]. The inset shows a cross section of the device with the mode field of the SPP in the gap. In addition, a lumped-element equivalent circuit of the PM is given. The device can be represented by a capacitor ($C_{\text{Device}} \approx 1.5\ldots3$ fF, length dependent). (b) Optical microscope photograph of the fabricated POH PM. Light is launched in and out of the silicon waveguide via diffraction grating couplers. The phase of the optical signal is modulated in the plasmonic modulator section having a length of 29 μm. (c) Optical microscope image of the fabricated MZM. The Mach-Zehnder interferometer (MZI) is fabricated on the silicon-on-insulator (SOI) platform using low-loss photonic MMI couplers for light splitting and combing. An optical path difference is implemented in the MZI, and the operating point is selected by choosing the operating wavelength. For modulation, an optical phase difference between the two arms is induced by the SPP phase shifters [122]. (d) Scanning electron microscope picture of the silicon-plasmonic MZM. The modes of the silicon waveguide are coupled to the plasmonic PM sections, which modulate the phase of the SPPs.

4.2.3 Plasmonic-organic Hybrid Phase Modulator

We first characterize the electro-optic response of the PM. Light from a tunable laser source (TLS) is coupled in and out of the silicon nanowires using the grating couplers. To characterize the modulator's radio frequency (RF) response, the device is driven with sinusoidal RF signals of various frequencies. The driving RF power is kept constant during the frequency sweep by first performing a power calibration before attaching the RF probe. The spectrum of the phase modulated optical signal $\cos(\omega_0 t + \eta \sin\omega_m t)$ is measured with an optical spectrum analyzer, and the modulation index η is estimated by the ratio of the power at the carrier wavelength and the power at the first sideband [122]. The measured modulation index as a function of the modulating RF frequency $f_m = \omega_m / 2\pi$ from $f_m = 10$ GHz up to $f_m = 62$ GHz is given in Figure 4.7(a) for an RF power of –10 dBm ($U_{pp} = 0.2$ V) measured across a 50 Ω resistor. In the measurements, the lower and the upper limits of the frequency range are posed by the resolution bandwidth of the optical spectrum analyzer and by the maximum frequency (for an RF power of –10 dBm) of the vector network analyzer, respectively. The black solid line in Figure 4.7(a) gives the mean value of two consecutive measurements, while the error bars describe the differrence between the two measurements. By knowing the RF response of the single PM, we estimate the 3 dB bandwidth to be larger than 60 GHz for a MZM comprising our PM and operating at the quadrature point of the MZM transfer function. The slight drop of the modulation index with increasing frequency is attributed to the RF probe, the frequency dependence of which could not be taken into account during the calibration of the RF power at the coaxial probe input. For clarity, we add the decaying frequency response of the probe as a grey solid line in Figure 4.7(a). In the range below 10 GHz, a flat frequency response is to be expected because of the small capacitance of the device, which has been demonstrated previously [122]. To demonstrate that the new EO material does not also limit the optical bandwidth of the device, we characterize the optical response of the BPSK modulator by keeping the RF constant at 40 GHz, and by sweeping the carrier wavelength across the 70 nm range around 1550 nm, see Figure 4.7(b). The flat optical response of the modulator makes the device also suitable for communication systems utilizing wavelength division multiplexing schemes.

For a modulation index of $\eta = 0.04$ at a sinusoidal RF modulation signal with 40 GHz and a peak-to-peak amplitude of $U_{pp} = 0.4$ V (open circuit condition), we estimate the value of the electro-optic coefficient to be $r_{33} = 116$ pm/V for a voltage–length product $U_\pi \times L$ of 0.45 Vmm. It has been previously shown that larger electro-optic coefficients and thus a smaller $U_\pi \times L$ can be achieved with an optimized poling [154].

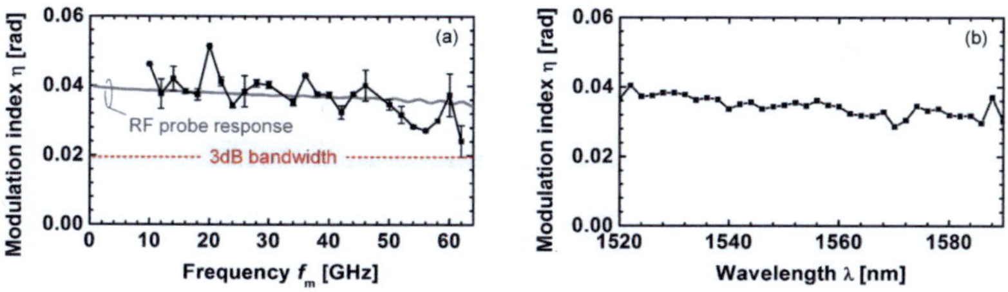

Figure 4.7 Electro-optic response of a phase modulator. (a) Modulation index as a function of the modulating frequency for a sinusoidal RF modulation signal. The RF power amounts to –10dBm measured for a 50 Ω terminating resistor. The black solid line gives the mean of two consecutive measurements, and the error bars indicate the difference between the two. For comparison, the grey solid line shows the transfer function of the GS probe from its coaxial input to the GS output as taken from the data sheet. The vertical scale is adjusted to show that for the higher frequencies the drop in modulation index is—to a significant amount—due to the frequency response of the probe. 3dB bandwidth of a MZM comprising the plasmonic PMs is indicated as a red dashed line. (b) Modulation index versus carrier wavelength in a wavelength range of 70 nm for the modulating RF $f_m = 40$ GHz.

Next, we encode a BPSK formatted data stream with a bitrate of 30 Gbit/s, 35 Gbit/s and 40 Gbit/s using the same experimental setup which was used in our previous work [122], see Figure 4.8(a). Light with a wavelength of 1550.92 nm is amplified in an erbium-doped fiber amplifier (EDFA) to a power level of +10 dBm and launched into the chip. The phase of the SPP is encoded with a 2^{31}–1 long pseudo-random bit sequence (PRBS) at a voltage swing of $U_{pp} = 3.8\ldots4.2$ V, measured across a 50 Ω resistor. The resulting BPSK signal is amplified and then detected by an Agilent N4391A Optical Modulation Analyzer (OMA). An equalizer with a filter length of 60 symbols is used. Constellation diagrams with the corresponding error vector magnitudes (EVM) are depicted in

Figure 4.8(b)-Figure 4.8(d) with 51,200 points per plot. The BER corresponding to the measured EVM of 18 % is below 10^{-10} [145], which is 5 orders of magnitude smaller than the previous demonstration [122]. EVMs measured without an equalizer are ~20 % i. e., the BER is less than 10^{-10} regardless of the equalizer. The low BER is made possible not by the digital signal processing, but by the large EO coefficient of YLD / PSLD, by the low optical losses in the silicon PLC, and by the comparatively low-loss plasmonic PM itself. Moreover, these low BERs are achieved with significantly lower optical and RF power levels than in the previous demonstration [122]. Representing the PS as a pure capacitor with an estimated capacitance of $C_{\text{device}} = 4.5$ fF, an energy consumption of $(2U_{\text{pp}})^2 \times C_{\text{device}} / 4 \approx 70$ fJ/ bit can be found for 40 Gbit/s operation [147].

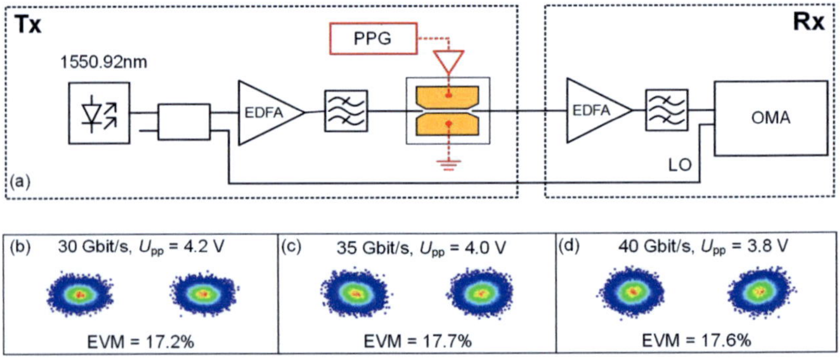

Figure 4.8 Experimental setup and constellation diagrams of BPSK data experiments performed with the plasmonic phase modulator. (a) Experimental setup used for generation and detection of the BPSK signals. (b), (c), (d) Constellation diagrams for the bit rates of 30 Gbit/s, 35 Gbit/s, 40 Gbit/s, respectively. Measured EVM are about 18 % corresponding to an BER smaller than 10^{-10} [145].

4.2.4 Plasmonic-organic Hybrid Mach-Zehnder Modulator

The POH MZMs are first characterized for their static electro-optic response. Light from a tunable laser source is coupled in and out of the silicon nanowire waveguide via diffraction grating couplers. Power transmission spectra for three MZMs with various PS lengths are given in Figure 4.9(a). In addition, we display the transmission spectrum of a reference MZI without a PS. It can be seen that the total insertion losses of the MZMs amount to 27…34 dB. According to the reference spectrum given in Figure 4.9(a), 14 dB of this loss are caused by the silicon grating couplers, MMIs, and waveguides. The on-chip insertion loss of the PS hence amounts to 13…20 dB, depending on the length of the PS. Variations in the slot widths and in the alignments of the metallic tapers can result in a power imbalance and in a change of the optical path length difference in the arms of the MZI, and therefore, in an extinction ratio and a free spectral range (FSR) that are different from the one of the reference MZI, see Figure 4.9(a). We analyze the shift of the wavelength corresponding to the minimum transmission as a function of the applied voltage in order to estimate the voltage U_π which is required for a phase shift of π. An example of the transmission spectrum for voltage-on and voltage-off states is given in Figure 4.9(b) for the MZM with 39 µm long phase modulator sections. Measuring the shift $\Delta\lambda_0$ for an applied voltage U_0, we can calculate with a help of the formula $U_\pi = \Delta\lambda_{FSR} \times U_0 / (2\Delta\lambda_0) \approx 30$ V for the MZM with 39 µm long PS, and an $U_\pi = 37$ V for the MZM with 29 µm long PS. The corresponding voltage-length product amounts to $U_\pi \times L \approx 1.1$ Vmm — one order of magnitude below the values of 10 Vmm as measured for silicon-based carrier-depletion modulators [155]. The in-device electro-optic coefficient r_{33} of the cladding amounts to 70 pm/V and is significantly lower than the maximum value of 110 pm/V specified for bulk SEO100. Thus, U_π values can be further improved by optimizing the poling procedure of the EO material.

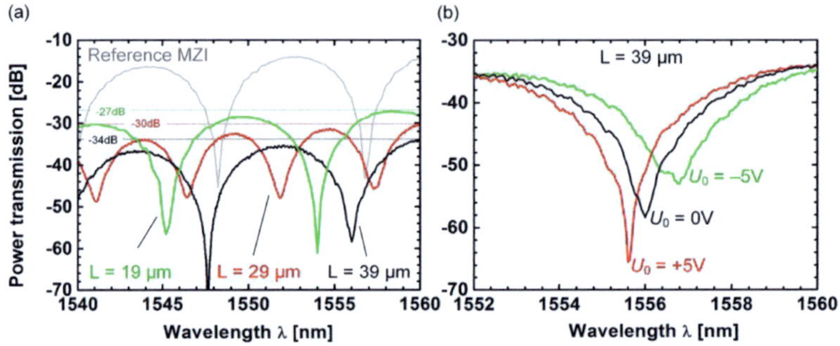

Figure 4.9 Static characterization of Mach-Zehnder modulators. (a) Fiber-to-fiber power transmission for MZMs with lengths of 19 µm, 29 µm and 39 µm. In addition, the transmission spectrum of a reference Mach-Zehnder interferometer without PS sections is shown. The plasmonic PS add 13 dB…20 dB additional optical loss. (b) Transmission spectrum of a 39 µm long device for various applied voltages U_0. Analyzing the shift of the wavelength corresponding to the minimum transmission we estimate the voltage U_π required for having a phase shift of π.

We perform data modulation experiments with our POH MZM using a direct receiver setup as shown in Figure 4.10(a). An electrical non-return-to-zero (NRZ) signal with PRBS pattern length of $2^{31}-1$ and with a peak-to-peak voltage swing of 5 V (measured across a 50 Ω resistor) is fed to the modulator via a ground-signal-ground (GSG) RF probe. The operating point for the MZM is defined by selecting the operating wavelength. The MZMs are operated in the quadrature points, i.e., the modulator output intensity changes linearly with the relative phase difference of the two arms. The OOK signal after the MZM is detected with a standard pre-amplified direct receiver comprising a single ED-FA, an optical band-pass filter with a bandwidth of 2 nm, a bit-error-ratio tester (BERT), and a digital communication analyzer (DCA).

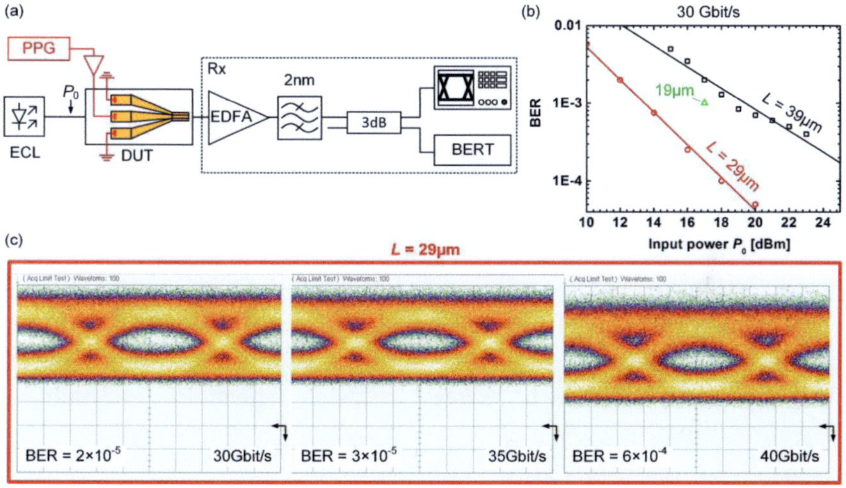

Figure 4.10 Modulation experiments with plasmonic silicon-organic MZMs with PS lengths of 19 µm, 29 µm and 39 µm. (a) Direct receiver setup used for detecting on-off keyed signal after the plasmonic MZMs. (b) Bit error ratios measured for the MZMs with plasmonic phase modulator sections having lengths of 19 µm, 29 µm and 39 µm. To find the optimum phase shifter length, we vary the input power to the modulators and measure the BER. A compromise between the optical loss and the modulation index is achieved by using a MZM with a PS length of 29 µm. (c) Eye diagrams measured at bit rates of 30 Gbit/s ($\mathrm{BER} = 2\times10^{-5}$), 35 Gbit/s ($\mathrm{BER} = 3\times10^{-5}$) and 40 Gbit/s ($\mathrm{BER} = 6\times10^{-4}$) for a MZM with 29 µm long PS sections at an input optical power of 20 dBm and at an operating wavelength of 1556.8 nm. The difference in the DC levels for data rates of 35 Gbit/s and 40 Gbit/s is attributed to the thermal drift of the operating point as a consequence of the large optical input power.

We measured the BERs for all three MZMs at a bit rate of 30 Gbit/s in order to find the optimum length for the phase modulators. During the experiment, the EDFA of the receiver is operated in constant output power mode. The input optical power to the modulator is varied from +10 dBm to +23 dBm. This varies the input power to the receiver, i.e., the optical signal-to-noise power ratio (OSNR) at the photodiodes. The optimum length of the PS is defined by a compromise between insertion loss and modulation index — making the device too short results in a small optical modulation amplitude, while a too long phase modulator section increases the insertion loss of the device, effectively decreasing the receiver's input power. We find that in our case ($U_{\mathrm{pp}} = 5$ V, SPP propagation losses of ~0.4 dB / µm, r_{33} = 70 pm/V) the optimum performance can be achieved with 29 µm long phase modulators, see Figure 4.10(b). A better BER can be achieved either by increasing the optimum PS length *L* by improving the slot quality (decreasing optical losses), or by reducing the effective PS length by increasing the electro-optical coefficient and reducing the slot width (increasing the optical modulation amplitude). The eye diagrams measured after the MZM with 29 µm long PS sections for bit rates of 30 Gbit/s (BER = 2×10^{-5}), 35 Gbit/s (BER = 3×10^{-5}) and 40 Gbit/s (BER = 6×10^{-4}) are given in Figure 4.10(c). These BER are well below the threshold of 4.5×10^{-3} for hard-decision FEC codes with 7% overhead [156]. The driving voltages and the optical insertion losses can be further reduced by, first, optimizing the poling procedure and thereby achieving higher electro-optic coefficients, second, reducing the slot size, and third, by using silver instead of gold.

4.2.5 Conclusion

We report on error-free generation and detection of 40 Gbit/s binary phase-shift keying (BPSK) signals using a 29 µm plasmonic-organic hybrid (POH) phase modulator with an energy consumption of 70 fJ/bit. Furthermore, we report on on-off keying (OOK) signaling with POH Mach-Zehnder modulators (MZM) at data rates of up to 40 Gbit/s with low energy consumptions of 75 … 225 fJ/bit. The reported MZM comprise POH phase modulators with lengths of 19 µm, 29 µm and 39 µm. Using the 29 µm device, we show OOK signaling at data rates of 30 Gbit/s, 35 Gbit/s and 40 Gbit/s with the BERs well below the hard-decision FEC threshold. The measured BERs represent the lowest values that have so far been shown with plasmonic modulators at technically relevant data rates. Our findings clearly demonstrate the applicability of plasmonic devices particularly in short-reach optical links.

[End of Paper]

4.3 SPP Absorption Modulator

The content of this section is a direct copy of the Journal publication [J4].

A. Melikyan, N. Lindenmann, S. Walheim, P. M. Leufke, S. Ulrich, J. Ye, P. Vincze, H. Hahn, Th. Schimmel, C. Koos, W. Freude, and J. Leuthold, "Surface plasmon polariton absorption modulator," Opt. Express, vol. 19, no 9, pp. 8855-8869, (2011).

Minor changes have been done to adjust the position of figures and notations of variables.

[Start of Paper]

Abstract: An electrically controlled ultra-compact surface plasmon polariton absorption modulator (SPPAM) is proposed. The device can be as small as a few micrometers depending on the required extinction ratio and the acceptable loss. The device allows for operation far beyond 100 Gbit/s, being only limited by *RC* time constants. The absorption modulator comprises a stack of metal/insulator/metal-oxide/metal layers, which support a strongly confined asymmetric surface plasmon polariton (SPP) in the 1.55 µm telecommunication wavelength window. Absorption modulation is achieved by electrically modulating the free carrier density in the intermediate metal-oxide layer. The concept is supported by proof-of-principle experiments.

4.3.1 Introduction

Optical modulators in future integrated circuits should enable data rates of 100 Gbit/s and beyond; they should be compact with footprints comparable to those of electronic devices (typically $< 1\,\mu m^2$), and fabrication with standard CMOS technology should be possible. Electro-absorption modulators (EAM) belong to the most compact commercially available data encoding devices in optical communications. They are available in typical lengths of 200 µm, supporting data rates of 40 Gbit/s [157]. Analog bandwidths can be as high as 100 GHz for traveling wave structures [158]. Unfortunately, high-speed EAM usually rely on expensive InP technology, which is challenging for economical mass production.

Besides EAMs, more recently, CMOS-compatible silicon-based solutions have been suggested [11, 90]. High speed all-silicon and silicon organic hybrid (SOH) modulators with lengths of a few millimeters and speeds exceeding 40 Gbit/s have already been demonstrated [18, 27, 33]. More compact Si-based modulators have been demonstrated in resonant structures. These devices exploit free-carrier dispersion either in a forward or reverse-biased *pn* junction [27, 30, 31, 33, 34, 40, 41, 44, 45, 140], allowing for operation at 10 Gbit/s. The operating speeds of such devices are ultimately limited by the finite mobility of free carriers. Thus, an ultra-fast, ultra-compact modulator based on a CMOS compatible technology has yet to be developed.

This can be achieved by using nano-plasmonic structures that enable strong confinement of both electrical and optical fields, thereby enhancing interaction with the underlying material and hence the modulation of the optical signal. Several approaches have been proposed for electro-optical manipulations of surface plasmon polaritons (SPPs). A technique proposed by Nikolajsen *et al.* [16] exploits the temperature-dependence of the refractive index in benzocyclobutene, an effect with limited operating speed. The device by Dionne *et al.* [37] — the so-called PlasMOStor device — exploits the carrier-induced refractive index change in a silicon layer. While this device works well, its speed is again limited by the electron mobility in the 170 nm thick active silicon layer. Since early in the 1980s, various approaches have been proposed for SPP modulation employing the linear electro-optic effect, e.g. in a polymer sandwiched between two metallic patches [26, 120, 159–164]. The feasibility of this concept is still to be investigated. Recently, a new modulation technique has been reported by our group [165] and independently by Feigenbaum *et al.* [54]. The technique relies on carrier modulation in a metal-oxide layer guiding a SPP. The approach of Feigenbaum *et al.* exploits the phase modulation of SPP in the visible wavelength range by a voltage-induced modulation of the carrier density in a bulk ITO layer. The group has reported ellipsometry measurements in the wavelength range 500 nm…800 nm that show significant changes of the voltage-dependent refractive index in bulk ITO layers that are transparent at these wavelengths. In contrast, our approach is based on a change of the absorption experienced by an SPP at 1.55 μm telecommunication wavelength. The absorption change is again caused by a voltage-induced carrier density modulation in ITO [165]. The advantage of our approach is that it has the capability of operating at telecommunication wavelengths, and that it is potentially free from carrier-related speed limitations.

In Section 4.2, we provide in-detail simulations and give first experimental evidence of the feasibility of an ultra-compact, CMOS compatible plasmonic SPPAM operated by directly modulating the electron carrier density in a metal-oxide layer. The suggested modulator operates at speeds that are only limited by the electronic drive circuits. To the best of our knowledge this is the first conceptual report of such an SPPAM supported by measurements. Unlike many prior plasmon modulator concepts, our approach is well suited for an operation in the near infrared and in particular in the telecommunication window around 1.3 μm and 1.55 μm, i.e. at a wavelength range where the plasmon resonances at the conventional metal/insulator interfaces are broad and therefore cannot be used for efficient switching.

The section is organized as follows: We introduce a design of the SPPAM compatible with the standard silicon-on-insulator (SOI) platform in Section 4.3.2. In Section 4.3.3, the dispersion relation (DR) for the suggested four-layer configuration is derived in general. Further, the properties of the carrier accumulation layer are calculated by means of the Thomas-Fermi approximation, and an in-depth study of the DR for a particular case is given in Section 4.3.4 and Section 4.3.5. Section 4.3.6 is devoted to the description of recent experimental results which validate the SPP absorption modulation concept.

4.3.2 Design

The configuration of the suggested SPPAM is given in Figure 4.11. It comprises a silicon strip waveguide on a buried silicon-oxide layer. The central plasmonic section consists of a dielectric and a thin metal-oxide layers sandwiched between two metal layers. The metal layers serve as SPP waveguides and as electrodes for applying an electric field across the dielectric.

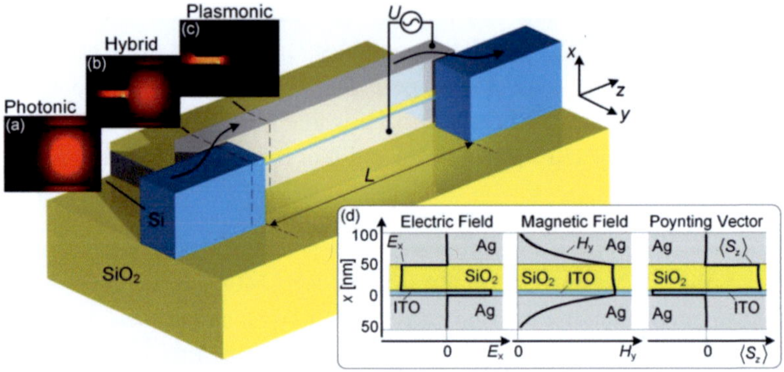

Figure 4.11 The structure of a surface plasmon polariton absorption modulator (SPPAM). Light is coupled from a silicon nanowire into an active plasmonic section by means of a directional coupler. The active section consists of a stack of silver (Ag), indium tin oxide (ITO), and SiO_2 layers. The absorption coefficient of the SPP is modulated by applying a voltage between the two silver electrodes. The insets show how a photonic mode (a) in a silicon strip waveguide excites a SPP (c) via a hybrid mode (b) in directional coupler. The insets in (d) show the electric field e_x and the magnetic field h_y as well as the time-averaged Poynting vector distributions $\langle S_z \rangle = \mathrm{Re}\{e \times h^*\}/2$ in the active plasmonic part. The plot of the Poynting vector shows the power confinement of the SPP in the ITO layer. The length L describes the size of the modulator along the light propagation direction.

The operation principle of the modulator is as follows. An optical quasi-TM mode featuring a dominant vertical electric field component is excited in the silicon strip waveguide (Figure 4.11, blue blocks). The light is subsequently coupled to the plasmonic waveguide, see the electric field magnitudes in Figure 4.11 insets (a)-(c). The illustrative numbers for the following description were taken from Table A.2 in Appendix A.10. Applying an electric field of, e. g., $10\,\mathrm{V}/30\,\mathrm{nm}$ changes the free carrier density in the intermediate metal-oxide layer (e.g., 8 nm thick ITO) from $N_0 = 9.25 \times 10^{26}\,\mathrm{m}^{-3}$ to $9.34 \times 10^{26}\,\mathrm{m}^{-3}$, thereby changing its complex permittivity from $\varepsilon = -1.667 + \mathrm{j}0.824$ to $-1.721 + \mathrm{j}0.832$. As a consequence, the absorption experienced by the SPP is decreased from $\alpha = 2.19\,\mu\mathrm{m}^{-1}$ to $2.07\,\mu\mathrm{m}^{-1}$. After propagating through the active plasmonic section of length L (e. g., 1 µm), the plasmonic mode couples back to the photonic mode inside the output coupling zone. As an overall result, the applied electric field modulates the output power.

In this work we have chosen for the active plasmonic section a sequence of four layers: From top to bottom (inset Figure 4.11) we have a silver (Ag) layer, a thin metal-oxide layer, here indium tin oxide (ITO), a silicon dioxide (SiO_2) film and silver layer. The whole structure is embedded in either PMMA or SiO_2. The choice of the dielectric sandwiched between two electrodes is based on studies of the dispersion relation introduced below. The device length can be in the range of a 1 μm up to a few micrometers, depending on the requirements with respect to the extinction ratio and the acceptable insertion loss.

The characteristics of ITO are key to the operation of the suggested structure. On the one hand ITO features a metallic behavior with relatively high carrier mobility, on the other hand, the carrier density is low to result in a small real part of the dielectric permittivity and a large electron screening length. This provides the following advantages for the operation of the modulator:

- There are no carrier-related speed limitations because of the high carrier mobility in ITO.
- The small magnitude of the complex permittivity of ITO compared to noble metals and typical dielectrics provides strong field enhancement within the ITO layer for the wavelength range of interest. Simulations in Figure 4.11 inset (d) show that the electric field in ITO is large, while there is hardly any field within the silver layer. If SiO_2 was replaced by Si_3N_4, the field confinement in the ITO layer could be further enhanced.
- The overlap between the externally induced charges and the SPP mode is particularly good in ITO compared to other metals because of the relatively large electron screening length. The large overlap is important to exploit the plasma-induced absorption change.
- Because in our case the SPP represents a field confined to an area far below the diffraction limit, a relatively small voltage creates an enormous electric field in the dielectric layer providing the necessary amount of carrier change in the ITO layer.

As a result of this design optimization, the size of the device can be as small as a few hundred nanometers in transverse direction by a few micrometers in length, a range which does not influence the modulator speed. Such techniques could allow integrating hundreds of modulators on the same chip.

4.3.3 SPP Dispersion Relation for a Four-layer Configuration

For discussing the basic operation principles of a plasmonic absorption modulator, we first investigate a 2D slab structure. We then identify key characteristic quantities, namely dispersion relation, insertion loss and extinction ratio, and study them in view of an actual 3D design. Finally, we specify the 3D SPP modal field and the propagation constant.

We derive here the dispersion relation (DR) of a four-layer active plasmonic structure, and we follow the procedure as described formerly [94, 95, 166]. The geometry of the problem and the respective permittivities are shown in Figure 4.12(a), (b). An SPP exist only in TM polarization, i.e. when the magnetic field is parallel to the metal-dielectric boundaries, see Figure 4.12. Therefore, we solve the wave equation for TM modes only.

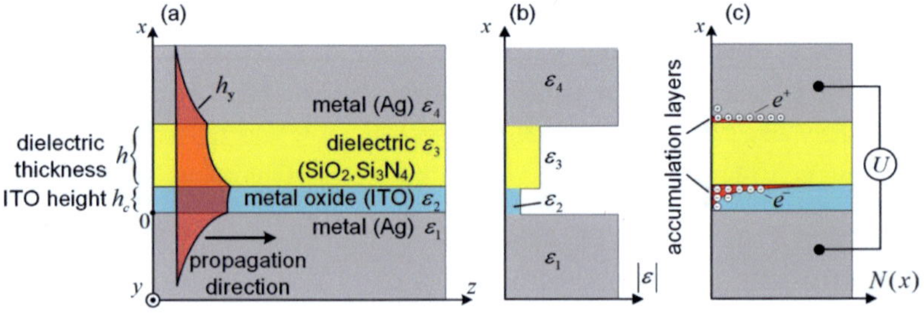

Figure 4.12 SPPAM with metal/dielectric/metal-oxide/metal layers. (a) Geometry and (b) dielectric permittivity distribution. The h_y component of the SPP magnetic field is schematically shown as a contour filled with reddish color in (a). The SPP propagates along the positive z-direction. (c) Carrier density distributions $N(x)$ in both electrodes

Assuming a time and z-coordinate dependency $e^{\mathrm{j}(\omega t-\beta z)}$ for the magnetic field h_y, the wave equation takes the form

$$\frac{\mathrm{d}^2\breve{h}_y(x,\omega)}{\mathrm{d}x^2}+\left[k_0^2\varepsilon(x,\omega)-\beta^2\right]\breve{h}_y(x,\omega)=0, \tag{4.1}$$

where $k_0=\omega/c$ is the magnitude of the wave vector of plane wave in free space, $\varepsilon(x)$ is the permittivity as a function of the x-coordinate and β is the propagation constant of the SPP. Without loss of generality we assume that the magnetic field h_y is confined in the dielectric and metal oxide layers and decays exponentially in the adjacent Ag layers. Thus, the solution of Eq. (4.1) satisfying the above requirements in each layer can be written with the transverse propagation constant $k_m=\sqrt{\varepsilon_m k_0^2-\beta^2}$ as

$$\begin{aligned}
&\breve{h}_y^1(x,\omega)=h_y^{1-}e^{\mathrm{j}k_1x} && \text{for} \quad -\infty<x<0\\
&\breve{h}_y^2(x,\omega)=h_y^{2+}e^{-\mathrm{j}k_2x}+h_y^{2-}e^{\mathrm{j}k_2x} && \text{for} \quad 0<x<h_c\\
&\breve{h}_y^3(x,\omega)=h_y^{3+}e^{-\mathrm{j}k_3(y-h_c)}+h_y^{3-}e^{\mathrm{j}k_3(y-h_c)} && \text{for} \quad h_c<x<h_c+h\\
&\breve{h}_y^4(x,\omega)=h_y^{4+}e^{-\mathrm{j}k_4(y-h-h_c)} && \text{for} \quad h_c+h<x<+\infty
\end{aligned} \tag{4.2}$$

The subscripts $m=1$ and $m=4$ are used for the bottom and top metal layers, the subscript $m=2$ denotes the metal oxide layer, and the subscript $m=3$ is assigned to the dielectric layer. The constants in front of the exponential functions describe the magnetic field amplitude in the respective medium. The electric field in each material can then be derived employing Maxwell's equations (ε_0 is the vacuum permittivity):

$$\begin{aligned}
e_z^m&=-\mathrm{j}\frac{1}{\omega\varepsilon_0\varepsilon_m}\frac{\partial h_y^m}{\partial x}\\
e_x^m&=-\frac{\beta}{\omega\varepsilon_0\varepsilon_m}h_y^m
\end{aligned} \tag{4.3}$$

Applying the boundary condition for the tangential components of the magnetic and the electric field allows to derive the dispersion relation

$$\frac{1+r_{12}\,\mathrm{e}^{-\mathrm{j}2k_2h_c}}{1-r_{12}\,\mathrm{e}^{-\mathrm{j}2k_2h_c}}=\frac{k_2\varepsilon_3}{k_3\varepsilon_2}\left[\frac{r_{34}\,\mathrm{e}^{-\mathrm{j}2k_3h}+1}{r_{34}\,\mathrm{e}^{-\mathrm{j}2k_3h}-1}\right], \tag{4.4}$$

Where $r_{12}=(k_2\varepsilon_1-k_1\varepsilon_2)/(k_2\varepsilon_1+k_1\varepsilon_2)$ and $r_{34}=(k_3\varepsilon_4-k_4\varepsilon_3)/(k_3\varepsilon_4+k_4\varepsilon_3)$. The solutions of Eq. (4.4) represent the eigenmodes of such a four-layer structure with their respective complex eigenvalues β. The imaginary part of β describes the absorption in the system, and the power absorption coefficient is defined as $\alpha=-2\,\mathrm{Im}[\beta]$.

The absorption modulator performance needs to be optimized for the lowest overall losses and the highest possible extinction ratio. Therefore it is convenient to introduce two parameters, namely the propagation length L_e and the length $L_{1\mathrm{dB}}$. The effective propagation length

$$L_e=\frac{1}{\alpha}=\left(-2\,\mathrm{Im}[\beta]\right)^{-1} \tag{4.5}$$

is the length after which the intensity of the SPP has decreased to a fraction of $1/\mathrm{e}$ of the initial intensity. In addition we define the 1dB *on-off* length $L_{1\mathrm{dB}}$ where the *on-off* ratio is ER = 1 dB for an (arbitrarily chosen) DC on-voltage of $U=10\,\mathrm{V}$ which is applied across the electrodes. The logarithmic extinction ratio ER is defined as

$$\begin{aligned}\mathrm{ER}&=10\lg\left(P_{\mathrm{on}}/P_{\mathrm{off}}\right)=10\lg\left(P_0\,\mathrm{e}^{-\alpha_{\mathrm{on}}L}/P_0\,\mathrm{e}^{-\alpha_{\mathrm{off}}L}\right)\\&=10\lg\left(\mathrm{e}^{|\alpha_{\mathrm{off}}-\alpha_{\mathrm{on}}|L}\right)=4.34\times|\alpha_{\mathrm{off}}-\alpha_{\mathrm{on}}|L\end{aligned} \tag{4.6}$$

The logarithmic extinction ratio depends linearly on the difference of the absorption coefficients $\Delta\alpha_{10\mathrm{V}}=|\alpha_{\mathrm{off}}-\alpha_{\mathrm{on}}|$ for the voltage-off and the voltage-on states. The relation between $L_{1\mathrm{dB}}$ and the 1 dB extinction ratio is

$$\mathrm{ER}_{1\mathrm{dB}}=1\mathrm{dB},\quad L_{1\mathrm{dB}}=\frac{1}{4.34\times\Delta\alpha_{10\mathrm{V}}}. \tag{4.7}$$

To calculate the absorption coefficient in the voltage-on and voltage-off states the Thomas–Fermi screening theory is employed.

Finally, we introduce a figure of merit (FoM) that relates the effective propagation length L_e to the 1 dB *on-off* length L_{1dB}. The FoM quantitatively describes the extinction ratio achievable in a device of length L_e operated with a 10 V voltage,

$$\mathrm{FoM} = L_e / L_{1dB}. \tag{4.8}$$

The larger the FoM, the better the performance is, i.e. the lower the propagation losses are for a 1 dB extinction ratio and an applied on-voltage of 10 V.

4.3.4 Thomas–Fermi Screening

An external voltage applied to the metallic electrodes changes the carrier density in the metal-oxide layer and hence influences the absorption coefficient of the SPP. For a proper estimation of this effect we employ the Thomas-Fermi screening theory for deriving the carrier distribution for a given voltage applied to the electrodes. The resulting spatial carrier distribution can then be used for modelling the complex dielectric permittivity of the metal-oxide layer via the Drude model, which then yields the complete information necessary for solving Eq. (4.4) for the on-state.

The Poisson equation relates the electric potential $\phi(x)$ to the induced charge density $N(x)$ within the ITO layer,

$$\Delta\phi(x) = \frac{e(N(x) - N_0)}{\varepsilon_0 \varepsilon_{\mathrm{ITO}}}. \tag{4.9}$$

Here, e is the elementary charge, ε_0 is the vacuum permittivity, $\varepsilon_{\mathrm{ITO}} = 9.3$ is the relative static permittivity of ITO [167], and N_0 is the free carrier density of ITO.

Following the Thomas-Fermi approach the total free carrier density and the potential are related by

$$N(x) = \frac{1}{3\pi^2}\left(\frac{2m_{\text{eff}}}{\hbar^2}\right)^{3/2}\left(E_F + e\phi(x)\right)^{3/2}, \tag{4.10}$$

where the Fermi energy is defined as

$$E_F = \left(\frac{\hbar^2}{2m_{\text{eff}}}\left[3\pi^2 N_0\right]^{2/3}\right). \tag{4.11}$$

In the latter expression $\hbar$ is Planck's constant. The electron effective mass m_{eff} for ITO is given in Table A.2 of Appendix A.10 in terms of the free electron mass $m_e = 9.1\times10^{-31}\,\text{kg}$. The free carrier density N_0 of ITO can be calculated from the plasma frequency ω_p in Table A.2 via the relation $N_0 = \varepsilon_0 m_{\text{eff}}\omega_\text{p}^2/e^2$.

In the Thomas-Fermi approximation, Eq. (4.9) and Eq. (4.10) have been solved analytically for the case of a weak potential ($e\phi << E_F$) resulting in the so-called Thomas–Fermi screening length $\lambda_{\text{TF}} = \left[\frac{\varepsilon_{\text{ITO}}\varepsilon_0\hbar^2}{m_{\text{eff}}e^2}\right]^{1/2}\left[\frac{\pi^4}{3N_0}\right]^{1/6}$, which gives an estimate of the carrier density distribution at the conductor interface. The Thomas-Fermi screening length for ITO is estimated to be $\lambda_{\text{TF}} \approx 6\,\text{Å}$, i.e. it is comparable to the value reported by Neumann *et al.* [167]. Obviously, due to the low free-carrier density of ITO, the estimated λ_{TF} is significantly larger for ITO than for noble metals where it is usually in the range of 1…2 Å.

Rather than employing the weak potential approximation, we numerically solve Eq. (4.9) with Eq. (4.10 - 4.11) for the ITO/dielectric interface with mixed boundary conditions

$$\frac{U}{h} = \varepsilon_{\text{ins}}\left.\frac{\mathrm{d}\phi}{\mathrm{d}x}\right|_{x\to +h_\text{c}} = \varepsilon_{\text{ITO}}\left.\frac{\mathrm{d}\phi}{\mathrm{d}x}\right|_{x\to -h_\text{c}}, \qquad \phi(x)\to 0 \quad \text{for} \quad x\to -\infty, \tag{4.12}$$

where U and d are the voltage and the distance between the two electrodes, respectively. The quantity ε_{ins} is the dielectric permittivity of the (insulating) dielectric. Due to their large dielectric strengths (high breakthrough voltage) silicon dioxide (SiO_2) with $\varepsilon_{SiO_2} = 3.9$ and silicon nitride (Si_3N_4) with $\varepsilon_{Si_3N_4} = 7.8$ have been considered as possible dielectrics for the above introduced structure. Inside the ITO layer, the second boundary condition is numerically approximated by taking the potential to be zero at a point many screening lengths λ_{TF} away from the ITO/dielectric interface.

Solving the boundary value problem provides the potential distribution $\phi(x)$ in the ITO layer. This is then used to calculate the induced carrier density $N(x)$ near the interface with the help of Eq. (4.10) and Eq. (4.11). In order to calculate the effect of this carrier density change on the DR it is more practical to use the average carrier density across the thin ITO layer with height h_c, $\langle N(x) \rangle_{h_c}$. This is justified by the fact that the SPP mode profile exhibits constant amplitude in the ITO layer, Figure 4.11. The relative change of the free carrier density in the 8 nm ITO layer is estimated to be 1 % for a 10 V voltage applied across a 30 nm thick SiO_2 layer. Below we show that this relatively small carrier density changes can be good enough for on-off switching.

4.3.5 SPP Dispersion Relation of Ag-ITO-dielectric-Ag

To study the effect of a carrier density change in the SPP guiding layers we have numerically solved the dispersion relation as derived in Eq. (4.4). The material dispersion of both ITO and silver are taken into account via the Drude model, see Section 2.1.2. In the following, a value of 2.19 PHz (0.46 PHz) is considered for the plasma frequency of Ag (ITO).

Dispersion relation and absorption spectrum of an multilayer structure comprising Ag / ITO(8nm) / Si_3N_4(70nm) / Ag (see inset Figure 4.13(b)) are plotted in Figure 4.13 as a function of frequency for two ITO carrier densities, N_0 and $\langle N_i(x) \rangle_{h_c}$. The quantities N_0 and $\langle N_i(x) \rangle_{h_c}$ correspond to densities without and with a voltage applied to the Ag-electrodes. For illustration, $\langle N_i(x) \rangle_{h_c}$ has been chosen to be $\langle N_i(x) \rangle_{h_c} = 1.05 \times N_0$. It can be seen that the dispersion of the SPP at the telecommunication frequency (193.5 THz) is basically not affected by carrier density changes within the ITO layer, Figure 4.13(a). However, the change of the absorption coefficient is as large as 30 %. Thus, the structure operates best as an absorption modulator rather than as a phase modulator.

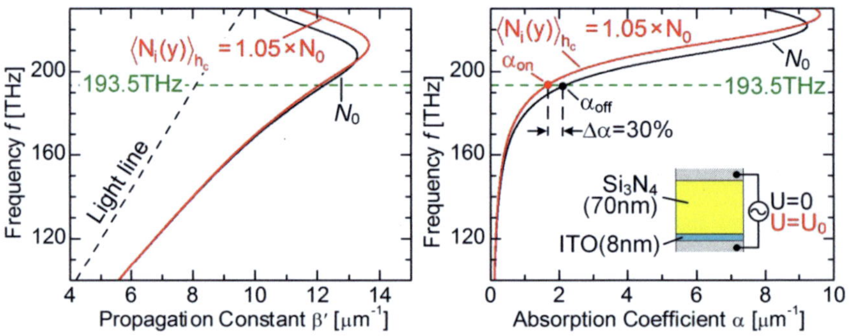

Figure 4.13 Dispersion relation of SPP guided by the Ag/ITO(8nm)/Si3N4(70nm)/Ag layer stack. Both, propagation constant and absorption coefficient, change when the carrier density of ITO is increased by 5 % (red lines). As opposed to the propagation constant the absorption coefficient varies significantly with ITO carrier density.

The maximum SPP propagation constant in Figure 4.13(a) defines a so-called surface plasma frequency which is close to the communication frequency at 193.5 THz due to the small plasma frequency of ITO. Decreasing the thickness d of the dielectric and increasing the ITO height h_c cause the maximum points in the dispersion relation Figure 4.13 to be shifted to the lower frequency region. As a consequence, the values of β' and α at the communication frequency 193.5 THz (green horizontal dashed line) increase. Bound SPPs at such a configuration exist to the right of the light line, i. e., for frequencies below 240 THz (or significantly above 240 THz [54] where the metallic property of ITO is less pronounced). The SPPs at the frequency range below 240 THz are well confined to the thin ITO layer because of its semi-metallic character. Therefore, the SPP absorption coefficient reacts very sensitively to a change of the free carrier density in ITO. This allows to efficiently modulate the intensity of light by electrically modulating the carrier density in the metal oxide layer.

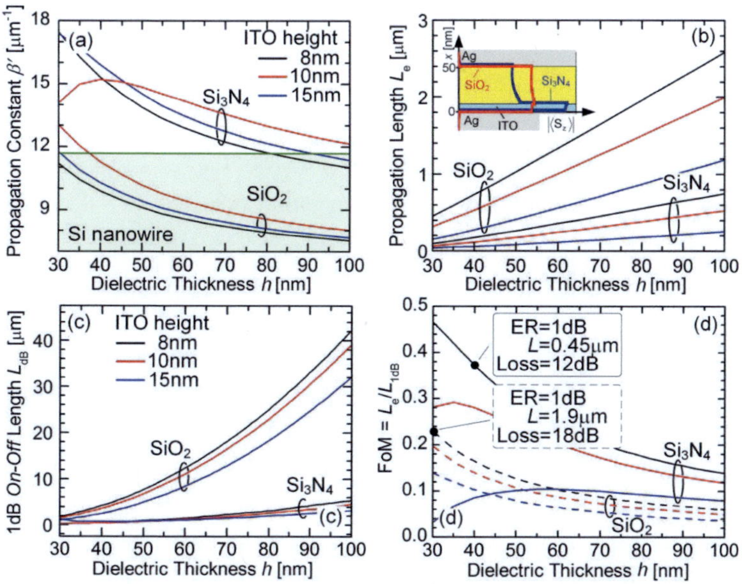

Figure 4.14 Dependence of characteristic device parameters as a function of the thickness h of the dielectric for two dielectric materials (Si_3N_4, SiO_2,) and for ITO thicknesses $h_c = 8$ nm (black line), 10 nm (red line) and 15nm (blue line). (a) Propagation constant β' . A typical propagation constant achievable in a silicon nanowire (500 nm × 500 nm) waveguide is shown as a horizontal green line. Therefore, an SPP excitation via a directional coupler is easily possible for a SiO_2 dielectric, but a grating coupler is preferable for a Si_3N_4 dielectric. (b) Propagation length L_e. Increasing the refractive index of the dielectric results in a decrease of L_e. The inset shows the absolute value of the time-averaged Poynting vector distributions in the structures filled with Si_3N_4 (blue line) and SiO_2 (red line). It can be seen that that the power confinement in ITO is larger than in the case of Si_3N_4. (c) Dependence of 1dB *on-off* length L_{1dB}. It can be seen that L_{1dB} is dramatically decreasing with the thickness of the dielectric, which is a result of both the static and the optical electrical field enhancement in the structure. The larger refractive index of the Si_3N_4 gives rise to a stronger optical field confinement in the active ITO layer, which in turn results in a shorter 1dB *on-off* length as well as shorter propagation length. (d) Figure of merit (FoM) as a function of the thickness of the dielectric. FoM increases for smaller dielectric thicknesses. The structure with Si_3N_4 dielectric performs considerably better. Insets give numerical examples for extinction ratio ER, device length L and loss in the system for both Si_3N_4 and SiO_2.

The modulator performance fundamentally depends on the thickness (h) and on the permittivity of the dielectric, and on the height (h_c) of the ITO. The dependencies of the propagation constant and absorption coefficient on h_c and h are depicted in Figure 4.14(a) and Figure 4.14(b) for two distinct dielectric materials, namely Si_3N_4 and SiO_2 with refractive indices of $\sqrt{\varepsilon_3} = 2.0$ and $\sqrt{\varepsilon_3} = 1.44$, respectively. It can be seen that the propagation constant β' decreases with increasing h. The propagation length L_e increases with h, because the portion of the optical power propagating in the lossy ITO layer decreases. The useful SPP propagation length does not exceed a few micrometers, which makes a proper choice of the ITO thickness crucial in the design. Further, one can see that L_e is significantly smaller for a Si_3N_4 than for a SiO_2 dielectric, which is a result of a stronger SPP field confinement in the ITO layer for the former case, see inset of Figure 4.14(b). The propagation constant β' as well as the propagation length L_e increases or decreases with increasing h depending on the ITO thickness and the refractive index of the dielectric layer, parameters which define the maxima in Figure 4.13(a) and Figure 4.13(b). For the same reason, extrema can be seen for the Si_3N_4-curves in Figure 4.14.

A significantly smaller 1 dB *on-off* length can be observed for the structure with a Si_3N_4 dielectric. Therefore, for finding an optimum with respect to propagation length and 1 dB *on-off* length as well as for a fair comparison of the structures with Si_3N_4 and SiO_2 dielectric, respectively, we display in Figure 4.14(d) the figure of merit from Eq. (4.8) in dependence of the dielectric and the ITO thicknesses. The graph in combination with the numerical values of the insets indicates that significantly improved extinction ratios with an acceptable loss can be achieved for a Si_3N_4 dielectric with device lengths less than 1 μm.

In order to make plasmonic modulators working devices, it is preferable to couple light from a photonic silicon wire waveguide into the modulation section, see Figure 4.11. However, the co-integration of plasmonics and photonics requires compact and efficient SPP excitation techniques. Several authors have proposed SPP excitation making use of a directional coupler configuration between the silicon and the plasmonic waveguide. Such coupling techniques require well-defined phase matching and a strong field overlap. It can be seen in

Figure 4.14(a) that phase matching is difficult with a Si_3N_4 layer, but can be easily achieved with a SiO_2 layer. In this case the SPP propagation constant is close to the one of a photonic quasi-TM mode in a Si nanowire, see the green colored area in Figure 4.14(a). Therefore, in spite of its large FoM for the Si_3N_4 dielectric, the subsequent discussions are based on a SiO_2 layer.

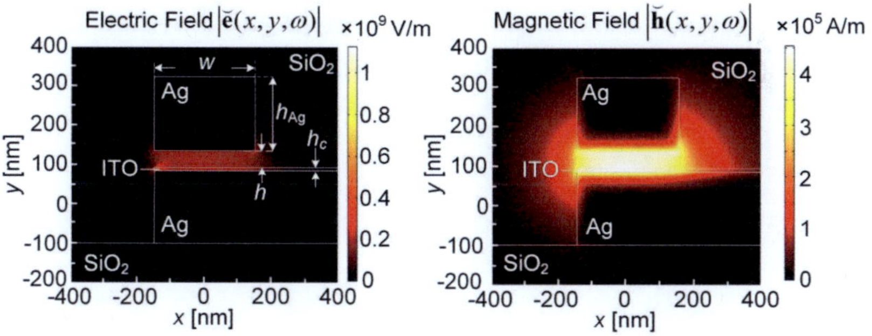

Figure 4.15 SPP mode profile in terms of dominating magnetic and electric field components at the active part of the modulator: Ag/SiO_2(40nm)/ITO(8nm)/Ag. The fields are normalized to a cross-section power of 1 W. Similar to the slab structure, h_c and h define the ITO and dielectric thicknesses, respectively. In addition, h_{Ag} and w describe the thickness and the width of the top silver electrode.

The two-dimensional (2D) model is helpful for the general understanding of the operation principles and for a first device optimization. In this section we perform a full vectorial eigenmode analysis of a realizable three-dimensional (3D) SPPAM structure as outlined in Figure 4.11. It consists of an Ag / ITO / SiO_2 / Ag multilayer structure on a SiO_2 substrate. To ensure a SPP confinement along the x-axis, the top Ag layer has a finite width w. It will be shown that the 2D results obtained previously are still valid for the 3D case.

The lateral cross-section of the plasmonic active part guiding the highly confined plasmonic mode is represented in Figure 4.15. The choice of the configuration is such that efficient SPP excitation is guaranteed and fabrication is feasible with state-of-the-art nanofabrication techniques. As in the 2D case, SPP of such a structure exhibit a large field confinement within the ITO layer (see

Figure 4.15). In addition to the dielectric and the ITO thickness some other geometrical 3D properties come into play, such as the width of the top silver strip w and the thickness of the silver layer h_{Ag} which is chosen sufficiently thick to avoid SPP coupling to the top side of the upper silver strip and to ensure that the dispersion relation derived in Section 4.3.3 is still valid. The eigenmodes of the 3D configuration have been simulated with COMSOL Multiphysics, see Appendix A.8. The simulations have been performed with rounded metal corners having a radius of curvature 4 nm. The propagation constant and length of the structure are plotted in Figure 4.16 with a comparison to the 2D slab version of the structure discussed in the previous subsection.

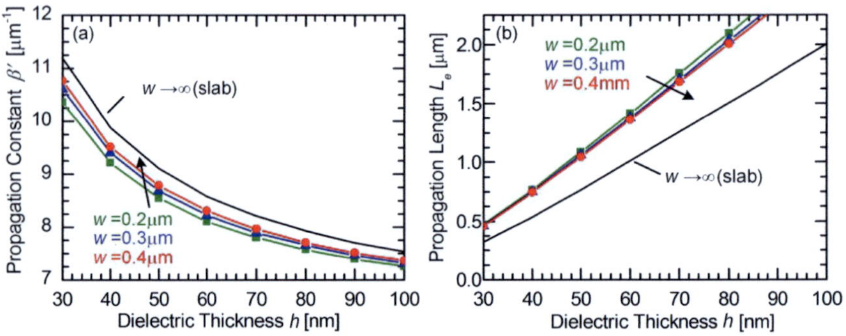

Figure 4.16 Influence of plasmonic 3D waveguide width w on propagation constant (a) and propagation length (b). The 2D results from solving Eq. (4.4) are included (black line).

Small deviations of the SPP propagation constant and propagation length in the case of the 3D structure are obviously caused by the finite values of w and h_{Ag}. Moreover, it can be seen that for relatively large values of w the propagation constant of SPP in the 3D structure approaches the one of the slab structure. As opposed to the results for the propagation constant much larger differences can be seen in the propagation length when going from the 2D slab structure to the 3D waveguides. The larger propagation length in the case of the 3D device is due to the decrease of the field confinement in the ITO. Therefore, one can assume that the required modulator length in a real device might be slightly longer than the values predicted by the 2D theory. However, both propagation constant and propagation length approach their 2D equivalents for small dielectric thicknesses. The latter is a consequence of a decrease of the h/w ratio. The SPP waveguide width w is an additional degree of freedom which can be used for manipulating both propagation constant and propagation length.

With a numerical 3D solution we have shown that the field can be horizontally confined along the x-axis under a silver bar with a finite width w, Figure 4.15. In such a three-dimensional wave guiding structure, all the requirements for an SPP modulation are fulfilled, namely high electric field strength in the dielectric layer, and optical field confinement in the ITO layer. The modulation speed is RC-limited, and estimate of this limiting frequency will be given next. The structure Figure 4.15 having a cross-section width of 300 nm and a length of 2μm has a capacitance of $C = 0.7\,\text{fF}$. Assuming an internal generator impedance of R = 50 Ω, this results in a typical RC time constant of $\tau = RC = 35\,\text{fs}$. Therefore, letting aside problems of impedance matching, the modulation bandwidth is expected to be in the THz-region.

4.3.6 Characterization

In a proof-of-principle experiment we performed modulation measurements using a structure similar to the one proposed in [165]. The modulator differs, however, from the one suggested in this reference in that the electrodes are 61 μm apart rather than 30…50 nm, see Figure 4.15. The large distance between the electrodes in this test device results in a very low electric field in the metal oxide layer and consequently in a small extinction ratio. However, the structure is simple to fabricate and enables us to give a first proof of principle.

A scanning electron microscope picture of the fabricated device before lifting off the silver (upper-left triangle outlined in green) is shown in Figure 4.17(b). For the fabrication of the device we used a standard silicon-on-insulator wafer with a 220 nm thick silicon layer on a 2 μm thick buried oxide. Silicon wire waveguides of 240 nm (and 400 nm) width and 220 nm height were fabricated using DUV lithography and standard CMOS processing. Silicon nitride with a 200 nm thickness was sputter-deposited. The ITO (10 nm) / Ag (250 nm) layers were then deposited on top of the Si_3N_4 layer via RF magnetron sputtering and electron beam evaporation, respectively. With photolithography the 10 μm long device was transversely structured along the silicon waveguides. A lift-off process removed the remaining silver. To decrease the spacing between the electrodes, the back of the wafer was partially thinned by wet etching with potassium hydroxide (KOH). In the etched-out area the silicon wafer was reduced to a thickness of ~ 60 μm. With electron beam evaporation a 500 nm thick electrical contact was then formed at the bottom of the silicon substrate.

Experiments were performed by applying an electrical signal with 10 V peak-to-peak amplitude and a rectangular shape. This modulation signal was provided by a function generator with variable frequency and an internal impedance of 50 Ω. TM polarized monochromatic light at a wavelength of 1.55 μm from a tunable laser source was coupled to the silicon strip waveguide via a lensed fibre. The resulting amplitude modulation of the optical signal was detected with an InGaAs photodetector connected to a lock-in amplifier. In order to subtract coherent electrical input-output crosstalk, the lock-in amplifier was calibrated with the DUT not being connected to the function generator.

Results of measured extinction ratios for modulation frequencies of up to 1.8 MHz are shown in Figure 4.17(c) for waveguides with 400 nm (■) and 240 nm widths (●), respectively. Measured extinction ratios are small but clearly detectable. The drop in the extinction ratio when increasing the modulation frequency is due to the low-pass characteristic of the device (RC limiting frequency ~58 kHz), see the equivalent circuit in the inset of Figure 4.17(a). The capacitor $C_1 = 1.9$ pF describes the layers between top metal and silicon substrate, while $C_2 = 0.2$ pF describes the silicon substrate itself. The silicon strip has such a large resistance that it can be represented by a capacitor. The capacitance values were calculated using the well-known dimensions and dielectric constants of the materials. The value of the resistance $R_2 = 1.2$ MΩ was calculated from the resistivity of the silicon wafer (material data provided by the manufacturer). With this model the voltage drop U across C_1 has been simulated with a conventional circuit simulator.

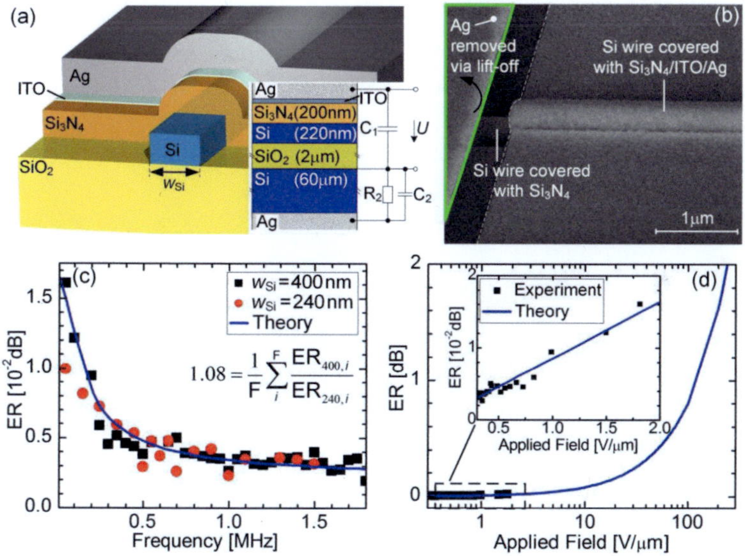

Figure 4.17 The modulator structure for a proof-of-principle experiment together with the measured extinction ratios versus modulation frequency. (a) 3D schematic of the fabricated device and its lumped element model describing the low-pass characteristic of the device. (b) Scanning electron microscope picture of the device before lift-off (the area taken inside the green contour has been removed after completing the lift-off process). The device length is $L = 10$ μm with an ITO thickness of 10 nm [165]. (c) Measured extinction ratios as a function of the frequency of the driving electrical signal. It can be seen that the extinction ratio does not depend on the silicon width w_{Si}. The extinction ratio calculated by the theory is represented as a blue line with the fit parameters $\omega_{\mathrm{p}} = 0.8 \times \omega_{\mathrm{p}0}$ and $\gamma = 2.3 \times \gamma_0$, where $\omega_{\mathrm{p}0}$ and γ_0 are plasma and collision frequencies of ITO from Table A.2. (d) Predicted ER for the strong electric fields obtained from the theory using fit parameter ω_{p} and γ.

This voltage U is then used for calculating the electric field in the Si_3N_4 layer and at the ITO / Si_3N_4 interface thereby defining the boundary condition Eq. (4.12) that is needed to solve Poisson's equation Eq. (4.9 - 4.11). The resulting induced carrier density N_i of the ITO layer enters the Drude model Eq. 2.14 together with the parameters of Table A.2. With $\varepsilon_{\mathrm{ITO}}$ of the ITO layer known, we employ the dispersion relation for a three-layer structure (equivalent to Eq. 4.4

for a four-layer device) and find the complex propagation constant β, the imaginary part of which gives the absorption coefficient α_{on} (α_{off} for $U = 0$). The difference $|\alpha_{\text{off}} - \alpha_{\text{on}}|$ defines the extinction ratio according to Eq. (4.6). In an iterative manner we modify ω_{p} and γ of Table A.2, until we reach a match for the measured ER.

A good match of the theoretically derived frequency response (blue line) and the experimental values (■,●) has been found for an ITO plasma frequency $\omega_{\text{p}} = 0.8 \times \omega_{\text{p0}}$ and a collision frequency $\gamma = 2.3 \times \gamma_0$, where ω_{p0} and γ_0 are published data [168], Table A.2. Our values for the ITO plasma frequency and collision frequency are red and blue shifted, respectively, relative to the values given in Table A.2. Such a red shift of the plasma frequency and a blue shift of the collision frequency has been previously reported for ultrathin ITO layers [169].

To prove that the detected small amplitude modulation is a result of the carrier modulation in the ITO film and that it does not originate from plasma or thermal effects in silicon, we performed additional measurements.

To exclude the potential influence of a plasma effect in silicon we measured the extinction ratios for silicon wire waveguides with two different widths w. Both silicon waveguides were covered with a 200 nm thick silicon nitride layer, see Figure 4.17(a). An eigenmode analysis revealed that the confinement factor of the quasi-TM mode is approximately two times larger for the 400 nm waveguide compared to the 240 nm waveguide. If the modulation was caused by the plasma effect in silicon, the measured extinction ratios for the wide and the narrow waveguide should be related by 2:1. According to Figure 4.17(c) and its formula inset this is not true: The average measured quotient of the two extinction ratios is 1.08, hardly larger than 1 and far away from 2.

Next, a potential thermal effect due to heating of the sample while applying a voltage has been considered. Heat could be dissipated in resistor R_2 of the equivalent circuit Figure 4.17(a). However, more heat would be dissipated at larger frequencies where a higher voltage drops across R_2 (as can be seen from the solid line in Figure 4.17(c) which shows that U decreases with frequency where the voltage between the silver electrodes is kept constant). As a consequence we would see a hypothetically larger extinction ratio at higher frequencies. However, according to our measurements in Figure 4.17(c) this is not true, so we exclude dominant thermal effects.

The final question to be answered is: What could be expected for the optimized structure Figure 4.11 where most of the voltage drops across the insulating SiO_2 layer? We plot the extinction ratio as a function of the modulating field strength as described before Figure 4.17. The inset Figure 4.17(d) shows the measured values (■) and the fitted curve (—). The main graph Figure 4.17(d) shows the ER prediction in a wider range of electric field strengths. A 1 dB extinction ratio can be obtained with an electric field of 100 V / μm applied to a 10 μm long device. Such an electric field strength can be achieved in the structure Figure 4.11. This field strength is far below the dielectric strength which is in the order of 10^3 V / μm for materials like SiO_2 and Si_3N_4.

4.3.7 Conclusion

We investigated an electro-optic surface plasmon polariton absorption modulator (SPPAM) operating at the 1.55 μm telecommunication wavelength. The underlying principles were analyzed. A three-layer prototype was designed, and first proof-of-principle experiments were performed. The device is *RC*-limited with a typical *RC* time constant of $\tau = 35\,\mathrm{fs}$. Therefore, the theoretical modulator speed is in the THz-region thus fulfilling future ultra-high speed data modulation requirements. For an optimized four-layer modulator with SiO_2 dielectric, a 1 dB extinction ratio has been predicted for a device length of 2 μm and a total plasmonic loss in the range of 18 dB. However, for the same extinction ratio of 1 dB the device length can be significantly decreased by utilizing Si_3N_4. With a modulation field strength of 250 V / μm the loss would be 12 dB for a device length of 0.5 μm which is comparable to the footprints of electronic devices.

[End of Paper]

5 Summary and Future Work

In the framework of this thesis, we developed plasmonic passive and active components for plasmonic integrated circuits in silicon photonic technology. The main objective of the thesis was to design and demonstrate ultra-compact plasmonic modulators in order to increase the integration density in future opto-electronic circuits.

Below, we summarize the key results and discuss the advances that can be implemented to improve the performance of the reported plasmonic devices.

5.1 Plasmonic Phase Modulators

We designed and fabricated novel ultra-compact plasmonic phase modulators (PPMs) based on the Pockels effect in electro-optic organic materials. In particular, we reported on a 29 μm long PPM integrated with silicon photonic circuit on the same chip. Generations of differential-phase-shift keying (DPSK) and binary-phase-shift keying (BPSK) optical signals at 40 Gbit/s have been demonstrated by employing the developed plasmonic modulators and bit error ratios (BERs) in the order of 10^{-5} have been measured. Comprehensive experimental analyses performed on a fabricated PPM have revealed its ultra-large RF and optical bandwidths of 65 GHz and 120 nm, respectively. In addition, we carried out temperature stability tests by measuring the phase modulation index for temperatures up to 85° C. The experiments showed that the reported PPMs have no performance degradations when increasing the operation temperature and, therefore, are well suited for applications in high temperature environments such as found in communication racks.

In the future, the usage of recent electro-optic materials with electro-optic coefficients of, e. g., 180 pm/V [76] in our device, will allow to decrease the drive voltage U significantly. Further reduction in the drive voltage can be achieved by decreasing the metallic slot width down to 70 nm, for instance. We therefore anticipate a decrease of U by at least a factor of 20 in the near future. For a device with a length of 21 μm and a slot width of 70 nm which is covered with recently developed electro-optic materials we predict a peak-to-peak phase modulation of $\pi / 2$ with a drive voltage swing of $U = 3$ V. With such a drive voltage swing and a device capacitance of 4 fF (based on the capacitance measured for the 1stG device) an average energy per bit of 9 fJ/bit can be anticipated. A $\pi / 2$ phase shift would also be needed to operate a MZM for generating advanced modulation formats. For an MZM the energy requirements then would be in the order of 18 fJ/bit. The optical loss can be decreased by making the device shorter through using narrower slots and better electro-optic organic materials. PPMs with smaller slots offer better performance, despite the increase of an attenuation constant with decreasing the slot width, as seen in Section 2.2.4. In particular, the smaller the slot width of a PPM is, the lower are the overall optical losses in the device, in spite of the increased attenuation constant of a gap SPP. The optical losses can be further reduced to 3 … 5dB by replacing the gold with silver, which provides lower optical losses for the SPP.

In Section 2.2.4 we showed that the modulation index achieved in a plasmonic phase modulator scales with the modulator length L. However, because of the high propagation losses of gap SPPs, the signal power at the output of the plasmonic modulator exponentially decreases with increasing device length. Therefore, an optimum device length should theoretically be found for a given transmitter power, receiver sensitivity, and drive voltage swing.

5.2 Surface Plasmon Polariton Absorption Modulator

We designed a novel electro-optic surface plasmon polariton absorption modulator (SPPAM) operating at the telecommunication wavelength of 1.55 µm. The modulator operates based on the plasma dispersion effect in an indium tin oxide layer (ITO). Induced carrier accumulation at the ITO / insulator interface shifts the absorption peak of a gap SPP. For an optimized four-layer modulator with SiO_2 dielectric, a 1 dB extinction ratio has been predicted for a device with a length of 2 µm and a total plasmonic loss in the range of 18 dB. However, for the same extinction ratio of 1 dB the device length can be even further decreased by utilizing Si_3N_4. The device is RC-limited with an estimated RC time constant of $\tau = 35\,\text{fs}$, i. e., with a cut-off frequency in the order of one THz. A prototype of the device was designed and fabricated on a silicon-on-insulator chip, and first proof-of-principle experiments were performed.

In the future, much higher extinction ratios can be achieved for the same drive voltage by reducing the insulator thickness down to 5 … 10 nm, which also has been proposed by other researchers [170]. The thinner the insulator thickness is, the larger will be the induced electron accumulation. This then results in a large shift of the absorption spectrum. Moreover, ITO films can be exploited in photonic-plasmonic hybrid structures, to modulate the attenuation constant of a hybrid photonic-plasmonic mode instead of a gap SPP [68]. Such hybrid plasmonic devices benefit from low propagation losses of hybrid modes and from a large carrier accumulation layer that can be induced with a small drive voltage applied across the gate oxide with a thickness of (5 … 10) nm.

5.3 Photonic-plasmonic Mode Converter

We designed a novel photonic-to-plasmonic mode converter for excitation of the gap SPP in a horizontal metallic slot waveguide (MSW) by the quasi-TM mode of a silicon strip waveguide. We implemented the eigenmode expansion method to optimize the geometrical parameters of the converter for its maximum conversion efficiency. Design parameters of the converter were reported for cladding materials with refractive indices of 1.44, 1.6 and 1.7. The converter provides more than 85 % conversion efficiency for metallic slot waveguides with slot heights of 30 … 50 nm.

Combining the developed mode converter concept with the concept of the horizontal MSW, a new generation of plasmonic phase modulators (PPMs) can be realized. PPMs with sub-50 nm slots can easily be fabricated by a standard "bottom-up" fabrication approach. This then can have a big impact on the energy consumption of the PPMs.

5.4 Plasmonic Polarization Beam Splitter

We have introduced a concept of a plasmonic polarization beam splitter employing the strong polarization sensitivity of metallic slot waveguides. We report on a plasmonic PBS with a length of 3.5 μm and with an extinction ratio of at least 20 dB. A device is designed for silicon photonic circuit and can be used to efficiently split two polarizations of silicon strip waveguides. The optical operation bandwidth of the reported PBS extends at least 30 nm.

Appendix

Appendix A. Theoretical Details

A.1. Fourier Transform of Real Functions

A time dependent physical quantity can equivalently be described in a frequency domain by means of its frequency components, which are related to the time dependent signal via the inverse Fourier transform

$$\tilde{\mathbf{A}}(\mathbf{r},\omega) = \int_{-\infty}^{+\infty} \mathbf{A}(\mathbf{r},t) e^{-\mathrm{j}\omega t} \mathrm{d}t,$$
$$\mathbf{A}(\mathbf{r},t) = \frac{1}{2\pi} \int_{-\infty}^{+\infty} \tilde{\mathbf{A}}(\mathbf{r},\omega) e^{\mathrm{j}\omega t} \mathrm{d}\omega. \tag{A.1}$$

Because the measurable physical quantities should be real, the relation for $\mathbf{A}\,(r,\,t)$ given in Eq. (A.1) can be modified to [82]

$$\mathbf{A}(\mathbf{r},t) = \frac{1}{2\pi} \int_{0}^{+\infty} \tilde{\mathbf{A}}(\mathbf{r},\omega) e^{\mathrm{j}\omega t} \mathrm{d}\omega + c.c. \tag{A.2}$$

where $c.c.$ states for a complex conjugate. In electromagnetics, complex amplitudes $\breve{\mathbf{A}}(\mathbf{r},\omega)$ of harmonic electromagnetic quantities $\mathbf{A}(\mathbf{r},t)$ are preferred over the Fourier transforms. If the spectrum of the time domain signal $\mathbf{A}(\mathbf{r},t)$ comprises discrete lines at frequencies ω_m where $m = 0, 1,\ldots M$, then we can write

$$\mathbf{A}(\mathbf{r},t) = \frac{1}{2} \sum_{m=0}^{M} \breve{\mathbf{A}}(\mathbf{r},\omega_m) e^{\mathrm{j}\omega_m t} + c.c. \tag{A.3}$$

A.2. Solution of the Vector Helmholtz Equation

The vector form of the Helmholtz equation is derived in Section 2.1.1 in the frequency domain which for isotropic materials can be written as

$$\nabla \times \nabla \times \breve{\mathbf{E}}(\mathbf{r},\omega) - \frac{\omega^2}{c^2} n^2(\mathbf{r},\omega)\breve{\mathbf{E}}(\mathbf{r},\omega) = 0. \tag{A.4}$$

If the refractive index $n(\mathbf{r},\omega)$ is translational invariant with respect to one of the coordinate axis, i. e., if

$$n(x,y,z,\omega) = n(x,y,z+r_0,\omega), \tag{A.5}$$

where r_0 is an arbitrary displacement along arbitrarily chosen axis z, then the solution of Eq. (A.4) can be shown to have the form [93]

$$\breve{\mathbf{E}}(\mathbf{r},\omega) = \breve{\mathbf{e}}(r_t,\omega)e^{-\mathrm{j}\beta z}. \tag{A.6}$$

where r_t stands for a position vector in transverse (x,y)-plane. The quantity β is called propagation constant and is a function of frequency. The propagation constant β defines the phase velocity of a monochromatic wave $\breve{\mathbf{E}}(r,\omega)$ at angular frequency ω. For a source-free homogenous linear medium we write the wave equation Eq. (A.4) [88]

$$\Delta\breve{\mathbf{E}}(\mathbf{r},\omega) + \frac{\omega^2}{c^2} n^2(\omega)\breve{\mathbf{E}}(\mathbf{r},\omega) = 0. \tag{A.7}$$

Thus, substituting the solution Eq. (A.6) into the wave equation Eq.(A.7) and assuming a plane wave in the transverse (x,y)-plane, the propagation constant β can be expressed in terms of the refractive index n [88]

$$\beta(\omega) = \frac{\omega}{c}\left[n'(\omega) - \mathrm{j}n''(\omega)\right]. \tag{A.8}$$

The negative imaginary part n'' of a refractive index describes an attenuation (or an amplification) of the wave amplitude defined by Eq. (A 6). Very often the power loss along the propagation direction z is characterized by a propagation length L_e after which the intensity of an electromagnetic wave decreases to a fraction of $1 / e$ its initial value,

$$L_e = -\frac{1}{2\,\mathrm{Im}(\beta)}. \tag{A.9}$$

Optical waveguides are usually translation-invariant with respect to the propagation direction, see Section 2.1.4. Therefore, the solution given in Eq. (A 6) also describes the wave which propagates in dielectric or plasmonic waveguides. In this case $\breve{\mathbf{e}}(r_t,\omega)$ is defined by a certain transverse resonant condition and describes the electric field profile of a mode with a propagation constant of β [93]. In general, $\breve{\mathbf{e}}(r_t,\omega)$ is a complex quantity. However, in the absence of the negative imaginary part of the refractive index n'', some of the transverse components of $\breve{\mathbf{e}}(r_t,\omega)$ are real [92]. For plasmonic waveguides where an imaginary part of the refractive index k cannot be neglected, the $\breve{\mathbf{e}}(r_t,\omega)$ is a tensor of rank one with non-zero complex components,

$$\breve{\mathbf{e}}(x,y,\omega) = \begin{pmatrix} \left|\breve{e}_x(r_t,\omega)\right| e^{-\mathrm{j}\phi_x(r_t,\omega)} \\ \left|\breve{e}_y(r_t,\omega)\right| e^{-\mathrm{j}\phi_y(r_t,\omega)} \\ \left|\breve{e}_z(r_t,\omega)\right| e^{-\mathrm{j}\phi_z(r_t,\omega)} \end{pmatrix}. \tag{A.10}$$

Thus, whenever giving an electric field profile of a plasmonic mode, the phase distribution $\phi_{x,y,z}(r_t,\omega)$ is needed in addition to the magnitude $\left|\breve{e}_{x,y,z}(r_t,\omega)\right|$.

A.3. Phase and Group Velocity

If an electromagnetic wave propagating in a waveguide has a limited spectral bandwidth of $\Delta\omega$, it is monochromatic if the condition $\Delta\omega/\omega \ll 1$ is fulfilled [88]. In this case, a spectral amplitude is non-zero only for the central frequency of ω_0 and an electric field in a time domain can be expressed with the help of Eq. (A.3) [88].

$$\mathbf{E}(\mathbf{r},t)=\frac{1}{2}\breve{\mathbf{e}}(r_t,\omega_0)e^{\mathrm{j}(\omega_0 t-\beta_0 z)}+c.c. \tag{A.11}$$

In Eq. (A.11), the $\breve{\mathbf{e}}(r_t,\omega_0)$ is the amplitude of a wave oscillating with a frequency of ω_0. It can be seen from the above expression that the phase of a monochromatic wave propagates with the so-called phase velocity,

$$v_p=\frac{\omega_0}{\beta_0}. \tag{A.12}$$

If an electromagnetic wave has a finite spectral bandwidth of Δω represents an optical pulse in a time domain. In this case, an electric field can be expressed in the time domain as [88]

$$\mathbf{E}(\mathbf{r},t)=\frac{1}{2\pi}\int_{-\infty}^{+\infty}\breve{\mathbf{e}}(r_t,\omega)e^{\mathrm{j}(\omega t-\beta z)}\mathrm{d}\omega=$$
$$=\frac{1}{2\pi}e^{\mathrm{j}\omega_0\left(t-\frac{\beta_0 z}{\omega_0}\right)}\int_{-\Delta\omega}^{+\Delta\omega}\breve{\mathbf{e}}(r_t,\omega'+\omega_0)e^{\mathrm{j}\omega'\left(t-\frac{\mathrm{d}\beta}{\mathrm{d}\omega}\Big|_{\omega=\omega_0}z\right)}\mathrm{d}\omega'+c.c., \tag{A.13}$$

where we expanded the frequency dependent propagation constant into a Taylor series and considered only the first term linear in ω. The phase of the optical carrier propagates with the phase velocity ω_0/β_0. The envelope of the modulated carrier propagates with the so-called group velocity

$$v_g=\left[\mathrm{Re}\left(\frac{\mathrm{d}\beta}{\mathrm{d}\omega}\Big|_{\omega=\omega_0}\right)\right]^{-1}. \tag{A.14}$$

For lossy materials such as metals, the propagation constant is complex. As a consequence, the imaginary part of the inverse derivative with respect to angular frequency distorts the envelope of the modulated wave [171]. Instead of the group velocity Eq. (A.14), a the group refractive index can be used,

$$n_g = \frac{c}{v_g}. \tag{A.15}$$

A.4. Poynting Vector and Power Flow

An electric field of a mode propagating in a dielectric or plasmonic waveguide can be given by Eq. (A.11). Similar relation also holds for a magnetic field $\mathbf{H}(r, t)$. Power flow which is given by the Poynting vector $\mathbf{S}(r, t)$ [81, 82] can have components along all three coordinate axis,

$$\mathbf{S}(r,t) = \mathbf{E}(r,t) \times \mathbf{H}(r,t). \tag{A.16}$$

For the time-averaged Poynting vector only the longitudinal component is of interest assuming that no power enters or leaks out of the waveguide [88],

$$\begin{aligned} \langle \mathbf{S}_z(r,t) \rangle = \langle [\mathbf{E} \times \mathbf{H}]_z \rangle \simeq \frac{1}{4} e^{-2\,\mathrm{Im}(\beta)z} \left[\breve{\mathbf{e}}(r_t,\omega) \times \breve{\mathbf{h}}^*(r_t,\omega) + \breve{\mathbf{e}}^*(r_t,\omega) \times \breve{\mathbf{h}}(r_t,\omega) \right]_z = \\ = \frac{1}{2} e^{-\alpha z} \, \mathrm{Re}\left\{ \left[\breve{\mathbf{e}}(r_t,\omega) \times \breve{\mathbf{h}}^*(r_t,\omega) \right]_z \right\}. \end{aligned} \tag{A.17}$$

The term $e^{-2\,\mathrm{Im}(\beta)z}$ describes the power attenuation along the propagation direction z, and can also be given in terms of the power attenuation constant α as $e^{-\alpha z}$. The attenuation constant α is often specified as a logarithmic measure,

$$\alpha_{\mathrm{dB}} = 10 \lg\left[1 / \mathrm{e}^{-\alpha z_0} \right] = 4.34 \alpha \mathrm{z}, \tag{A.18}$$

where z_0 is the reference propagation length.

A.5. General Orthogonality Relation

Following a routine similar to the one given in [92], here we derive the general orthogonality relation which also holds for plasmonic waveguides. The eigenmodes given by Eq. (A.6) describe the forward propagating modes which decay along the positive z direction in the case of lossy waveguides. Denoting the fields of the forward propagating fields with a (+) sign we write Faraday's and Ampere's laws given in Eq. (2.1) in the frequency domain as

$$\begin{aligned} \nabla \times \breve{\mathbf{E}}_{\mu}^{+}(\mathbf{r},\omega) &= -\mathrm{j}\omega\mu_{0}\breve{\mathbf{H}}_{\mu}^{+}(\mathbf{r},\omega) \\ \nabla \times \breve{\mathbf{H}}_{\mu}^{+}(\mathbf{r},\omega) &= \mathrm{j}\omega\varepsilon_{0}\underline{\varepsilon}(\mathbf{r},\omega)\breve{\mathbf{E}}_{\mu}^{+}(\mathbf{r},\omega) \end{aligned} \tag{A.19}$$

where μ stands for certain eigenmodes with a propagation constant β_{μ}. Multiplying both sides of the first equation with a $\breve{\mathbf{H}}_{\nu}^{-}$ and the second equation with $\breve{\mathbf{E}}_{\nu}^{-}$ and using the vector identity $\nabla(\mathbf{A}\times\mathbf{B})=\mathbf{B}(\nabla\times\mathbf{A})-\mathbf{A}(\nabla\times\mathbf{B})$ we write

$$\begin{aligned} &\nabla\left(\breve{\mathbf{E}}_{\mu}^{+}(\mathbf{r},\omega)\times\breve{\mathbf{H}}_{\nu}^{-}(\mathbf{r},\omega)\right)= \\ &\quad=-\mathrm{j}\omega\mu_{0}\breve{\mathbf{H}}_{\nu}^{-}(\mathbf{r},\omega)\breve{\mathbf{H}}_{\mu}^{+}(\mathbf{r},\omega)-\mathrm{j}\omega\varepsilon_{0}\breve{\mathbf{E}}_{\nu}^{-}(\mathbf{r},\omega)\underline{\varepsilon}(\mathbf{r},\omega)\breve{\mathbf{E}}_{\mu}^{+}(\mathbf{r},\omega), \end{aligned} \tag{A.20}$$

where we denoted the backward propagating fields with a minus sign. The multiplication sequence of the backward and forward propagating modes in Eq. (A.20) is chosen arbitrarily, therefore, we can also write

$$\begin{aligned} &\nabla\left(\breve{\mathbf{E}}_{\mu}^{-}(\mathbf{r},\omega)\times\breve{\mathbf{H}}_{\nu}^{+}(\mathbf{r},\omega)\right)= \\ &\quad=-\mathrm{j}\omega\mu_{0}\breve{\mathbf{H}}_{\nu}^{+}(\mathbf{r},\omega)\breve{\mathbf{H}}_{\mu}^{-}(\mathbf{r},\omega)-\mathrm{j}\omega\varepsilon_{0}\breve{\mathbf{E}}_{\nu}^{+}(\mathbf{r},\omega)\underline{\varepsilon}(\mathbf{r},\omega)\breve{\mathbf{E}}_{\mu}^{-}(\mathbf{r},\omega). \end{aligned} \tag{A.21}$$

Subtracting both sides of Eq. (A.21) from Eq. (A.20) we get

$$\nabla\left(\breve{\mathbf{E}}_{\mu}^{+}(\mathbf{r},\omega)\times\breve{\mathbf{H}}_{\nu}^{-}(\mathbf{r},\omega)-\breve{\mathbf{E}}_{\nu}^{-}(\mathbf{r},\omega)\times\breve{\mathbf{H}}_{\mu}^{+}(\mathbf{r},\omega)\right)=0. \tag{A.22}$$

Next, we use the two dimensional divergence theorem in Eq. (37–56) of Ref [92]

$$\int_{A}\nabla\mathbf{F}\mathrm{d}x\mathrm{d}y=\frac{\partial}{\partial z}\int_{A}F_{z}\mathrm{d}x\mathrm{d}y+\oint_{l_A}\mathbf{F}_{t}\mathrm{d}l. \tag{A.23}$$

The integration area A represents a planar surface perpendicular to the z-axis and having a perimeter l_{M} [92]. We take a surface integral of the LHS of Eq. (A.22) and use the identity Eq. (A.23) in order to further simplify Eq. (A.22). By selecting a sufficiently large integration area A, the line integral of Eq. (A.23) can be made zero because the electric and the magnetic fields of the bound modes decay in (x, y)-plane away from the core. This then results

$$\mathrm{j}\left(\beta_\mu^+ + \beta_\nu^-\right)e^{\mathrm{j}\left(\beta_\mu^+ + \beta_\nu^-\right)z} \int_A \left[\breve{\mathbf{e}}_\mu^+(r_t,\omega)\times\breve{\mathbf{h}}_\nu^-(r_t,\omega) - \breve{\mathbf{e}}_\nu^-(r_t,\omega)\times\breve{\mathbf{h}}_\mu^+(r_t,\omega)\right]_z \mathrm{d}x\mathrm{d}y = 0. \tag{A.24}$$

The propagation constant of the νth backwards propagating mode is denoted as β_ν^-. The fields and the propagation constant of a backward propagating mode are related to the ones of a forward propagating mode through a relation Eq. (11-9) of Ref [92]

$$\beta^+ = -\beta^-, \quad \breve{\mathbf{e}}^- = -\breve{e}_x^+\mathbf{x} - \breve{e}_y^+\mathbf{y} + \breve{e}_z^+\mathbf{z}, \quad \breve{\mathbf{h}}^- = \breve{h}_x^+\mathbf{x} + \breve{h}_y^+\mathbf{y} - \breve{h}_z^+\mathbf{z}. \tag{A.25}$$

Here, the vectors **x**, **y**, **z** represent the basis of the right-handed Cartesian coordinate system. We can rewrite Eq. (A.24) in terms of the quantities of a forward propagating mode as

$$\left(\beta_\mu^+ - \beta_\nu^+\right)\int_A \left[\breve{\mathbf{e}}_\mu^+(r_t,\omega)\times\breve{\mathbf{h}}_\nu^+(r_t,\omega) + \breve{\mathbf{e}}_\nu^+(r_t,\omega)\times\breve{\mathbf{h}}_\mu^+(r_t,\omega)\right]_z \mathrm{d}x\mathrm{d}y = 0, \tag{A.26}$$

The integral in Eq. (A.24) is non-zero only if $\nu \neq \mu$, and therefore, $\beta_\mu^+ \neq -\beta_\nu^-$.

Thus, the general orthogonality relation can be obtained from Eq. (A.24) and Eq. (A.25) by,

$$\begin{aligned}\int_A \left[\breve{\mathbf{e}}_\mu^+(r_t,\omega)\times\breve{\mathbf{h}}_\nu^+(r_t,\omega) + \breve{\mathbf{e}}_\nu^+(r_t,\omega)\times\breve{\mathbf{h}}_\mu^+(r_t,\omega)\right]_z \mathrm{d}x\mathrm{d}y = \\ = \delta_{\mu\nu} 2\int_A \left[\breve{\mathbf{e}}_\mu^+(r_t,\omega)\times\breve{\mathbf{h}}_\nu^+(r_t,\omega)\right]_z \mathrm{d}x\mathrm{d}y,\end{aligned} \tag{A.27}$$

where $\delta_{\mu\nu}$ is Kronecker delta function. Eq. (A.27) is the general orthogonality relation which holds for any waveguide, especially for the lossy modes of plasmonic waveguides.

A.6. Coupled-mode Theory

Field distributions $\breve{\mathbf{e}}_{in}(x,y)$ and $\breve{\mathbf{h}}_{in}(x,y)$ in a waveguide can be expanded in terms of the eigenmodes of the waveguide [92]

$$\begin{aligned}\breve{\mathbf{e}}_{in}(r_t,\omega) &= \sum_{\mu} c_{\mu}^{in}\,\breve{\mathbf{e}}_{\mu}(r_t,\omega)\\ \breve{\mathbf{h}}_{in}(r_t,\omega) &= \sum_{\mu} c_{\mu}^{in}\,\breve{\mathbf{h}}_{\mu}(r_t,\omega)\end{aligned} \tag{A.28}$$

Multiplying both sides of the first equation of Eq. (A.28) with $\breve{\mathbf{h}}_{\nu}$ and the second equation with $\breve{\mathbf{e}}_{\nu}$, taking the surface integral of both

$$\begin{aligned}\int_A \left[\breve{\mathbf{h}}_{\nu}(r_t,\omega)\times\breve{\mathbf{e}}_{in}(r_t,\omega)\right]_z \mathrm{d}x\mathrm{d}y &= c_{\nu}^{in}\int_A \left[\breve{\mathbf{h}}_{\nu}(r_t,\omega)\times\breve{\mathbf{e}}_{\nu}(r_t,\omega)\right]_z \mathrm{d}x\mathrm{d}y\\ \int_A \left[\breve{\mathbf{e}}_{\nu}(r_t,\omega)\times\breve{\mathbf{h}}_{in}(r_t,\omega)\right]_z \mathrm{d}x\mathrm{d}y &= c_{\nu}^{in}\int_A \left[\breve{\mathbf{e}}_{\nu}(r_t,\omega)\times\breve{\mathbf{h}}_{\nu}(r_t,\omega)\right]_z \mathrm{d}x\mathrm{d}y\end{aligned} \tag{A.29}$$

where we used the orthogonality relation Eq. (A.27) to simplify the RHS of the later equation. The relation for the coupling strength c_{ν}^{in} can be found by adding the second equation of Eq. (A.29) to the first one

$$c_{\nu}^{in} = \frac{1}{2}\frac{\int_A \left[\breve{\mathbf{e}}_{in}(r_t,\omega)\times\breve{\mathbf{h}}_{\nu}(r_t,\omega)+\breve{\mathbf{e}}_{\nu}(r_t,\omega)\times\breve{\mathbf{h}}_{in}(r_t,\omega)\right]_z \mathrm{d}x\mathrm{d}y}{\int_A \left[\breve{\mathbf{e}}_{\nu}(r_t,\omega)\times\breve{\mathbf{h}}_{\nu}(r_t,\omega)\right]_z \mathrm{d}x\mathrm{d}y} \tag{A.30}$$

More generally normalized coupling strength can be defined which is independent of the exact field amplitudes of the modes.

$$c_{\nu}^{in} = \frac{1}{2}\frac{\int_A \left[\breve{\mathbf{e}}_{in}(r_t,\omega)\times\breve{\mathbf{h}}_{\nu}(r_t,\omega)+\breve{\mathbf{e}}_{\nu}(r_t,\omega)\times\breve{\mathbf{h}}_{in}(r_t,\omega)\right]\mathrm{d}x\mathrm{d}y}{\sqrt{\int_A \left[\breve{\mathbf{e}}_{in}(r_t,\omega)\times\breve{\mathbf{h}}_{in}(r_t,\omega)\right]\mathrm{d}x\mathrm{d}y}\sqrt{\int_A \left[\breve{\mathbf{e}}_{\nu}(r_t,\omega)\times\breve{\mathbf{h}}_{\nu}(r_t,\omega)\right]\mathrm{d}x\mathrm{d}y}} \tag{A.31}$$

A.7. First-order Perturbation Theory

In deriving the orthogonality relation for lossy waveguides above, we have assumed the μ and ν in Eq. (A.20) correspond to two various modes of the same waveguide. Here, instead of the mode ν, we take a backward propagating mode μ'_{-} of a waveguide that has slightly perturbed material property ε' we can rewrite Eq. (A.20) as

$$\begin{aligned}&\nabla\left(\breve{\mathbf{E}}_{\mu}^{+}(r,\omega)\times\breve{\mathbf{H}}_{\mu'_{-}}^{\prime-}(r,\omega)\right)=\\&\quad=-\mathrm{j}\omega\mu_0\breve{\mathbf{H}}_{\mu'_{-}}^{\prime-}(r,\omega)\breve{\mathbf{H}}_{\mu}^{+}(r,\omega)-\mathrm{j}\omega\varepsilon_0\breve{\mathbf{E}}_{\mu'_{-}}^{\prime-}(r,\omega)\underline{\varepsilon}'(r,\omega)\breve{\mathbf{E}}_{\mu}^{+}(r,\omega)\end{aligned}\tag{A.32}$$

Because the sequence of the backward and forward propagating modes in Eq. (A.32) is chosen arbitrarily, a similar equation holds for flipped mode sequences. After subtracting the second one form the first one we get

$$\begin{aligned}&\nabla\left(\breve{\mathbf{E}}_{\mu}^{+}(r,\omega)\times\breve{\mathbf{H}}_{\mu'_{-}}^{\prime-}(r,\omega)-\breve{\mathbf{E}}_{\mu'_{-}}^{\prime-}(r,\omega)\times\breve{\mathbf{H}}_{\mu}^{+}(r,\omega)\right)=\\&\quad=-\mathrm{j}\omega\varepsilon_0\breve{\mathbf{E}}_{\mu_{-}}^{\prime-}(r,\omega)\underline{\varepsilon}'(r,\omega)\breve{\mathbf{E}}_{\mu}^{+}(r,\omega)+\mathrm{j}\omega\varepsilon_0\breve{\mathbf{E}}_{\mu}^{+}(r,\omega)\underline{\varepsilon}(r,\omega)\breve{\mathbf{E}}_{\mu'_{-}}^{\prime-}(r,\omega)\end{aligned}\tag{A.33}$$

Using the divergence theorem Eq. (A.23) and the relation between the backward and forward propagating modes Eq. (A.25), a change of a complex propagation constant can be expressed similar to the Eq. (18–62) in Ref. [92]

$$\Delta\beta=\beta'-\beta=\frac{k_0 2n_{xx}\Delta n_{xx}}{Z_0}\frac{\displaystyle\int_{\text{Active area}}\breve{\mathbf{e}}_x(r_t,\omega)\breve{\mathbf{e}}_x(r_t,\omega)\,\mathrm{d}x\mathrm{d}y}{\displaystyle\int_A\left[\breve{\mathbf{e}}(r_t,\omega)\times\breve{\mathbf{h}}(r_t,\omega)+\breve{\mathbf{e}}(r_t,\omega)\times\breve{\mathbf{h}}(r_t,\omega)\right]_z\mathrm{d}x\mathrm{d}y}\tag{A.34}$$

where $k_0 = 2\pi / \lambda$ is the wave vector of light in vacuum at a wavelength of λ, and Z_0 as the vacuum field impedance. While deriving Eq. (A.34) the following assumptions are made

- In the first order perturbation approximation, the field profiles of a mode of a perturbed waveguide are assumed to be the same as for the non-perturbed waveguide.
- Only ε_{xx} coefficient of the relative permittivity tensor of a lossless medium is changed. This is the case for electro-optic organic materials whenever poling and modulating electric fields are oriented along the x-axis, see Section 2.1.3.
- The second-order change in the refractive index is neglected. The permittivity change is expressed in terms of the refractive index change Δn_{xx} via $\Delta\varepsilon_{xx} = 2n_{xx}\Delta n_{xx}$, where the n_{xx} and Δn_{xx} are the refractive index and the refractive index change along the x-axis, respectively.

In addition, we have omitted the sign (+) in Eq. (A.34) in order to keep the equation in a more general form.

A.8. Simulation Methods

Several simulation tools have been used throughout this thesis:

COMSOL Multiphysics [126]

We have used COMSOL Multiphysics (CM) for calculating the eigenmodes of all the waveguides. COMSOL uses the finite element method [172] to solve Maxwell's equations both in time and in frequency domain. It has an in-build mode analyzer which is capable of finding both real and complex eigenvalues of any lossless or lossy waveguides. COMSOL uses a tetrahedral mesh grid, and therefore is suitable for modelling arbitrary three-diemensional metallic structures. This is needed for properly approximating the corners of the metals which are designed to have a curvature with a radius of (10 … 20) nm. Parameter sweeps and post processing of the results generated by COMSOL is done in MATLAB. This is the preferred procedure whenever numerical integration and fitting of various quantities produced by COMSOL is needed.

RSOFT [127]

RSoft is an electromagnetic simulation tool comprising several modules which can be used to solve various electromagnetic design problems.

FullWAVE uses a finite-difference time-domain (FDTD) method to solve Maxwell's equations in the time domain. Using FullWAVE with a temporal impulse excitation, transmission and reflection spectra of the device can be simulated by performing a fast Fourier transform (FFT) on the transmitted or reflected time domain signals. A detailed interpretation of the calculated spectra can be obtained by using a continuous wave (CW) excitation and analyzing the electric field distribution in the device at a single wavelength.

FEMSim is an electromagnetic mode solver based on the finite element method. We used FEMSim to calculate a mode of the silicon waveguide which is then used in FullWAVE as an excitation signal for impulse or CW excitation.

CST Microwave Studio [128]

CST Microwave Studio (MWS) solves Maxwell's equations in integral form. CST-MWS can be used for solving an electromagnetic problem in both time (TD) and frequency (FD) domain. We used the CST-MWS time domain solver to study light propagation in our devices.

A.9. Dispersion Relation of Multilayer Waveguides

We implemented a numerical model which calculates the dispersion relation for an SPP in an arbitrary multilayer slab structure following the matrix method [173]. We assume that N layers of arbitrary material, but homogeneously along the z-axis, are stacked along the x-direction. We then derive the wave equation in each layer for the transverse magnetic (TM) mode $\breve{\mathbf{H}}(r,\omega) = \breve{h}_y(x,\omega)e^{-\mathrm{j}\beta z}$ and transform the wave equation into the frequency domain Eq. (A.7) [1],

$$\frac{\partial^2 \breve{h}_y^{(m)}(x,\omega)}{\partial x^2}+(k_0{}^2 n_m^2-\beta^2)\breve{h}_y^{(m)}(x,\omega)=0,\cdot \tag{A.35}$$

The quantity β is the propagation constant, n_m is the complex refractive index of the mth layer, and $k_0=\omega/c$ is the wave vector of light in vacuum. We also used the fact that for TM modes the magnetic field components $\breve{h}_x^{(m)}$ and $\breve{h}_z^{(m)}$ are zero as well as $\partial\breve{h}_y^{(m)}/\partial y=0$.

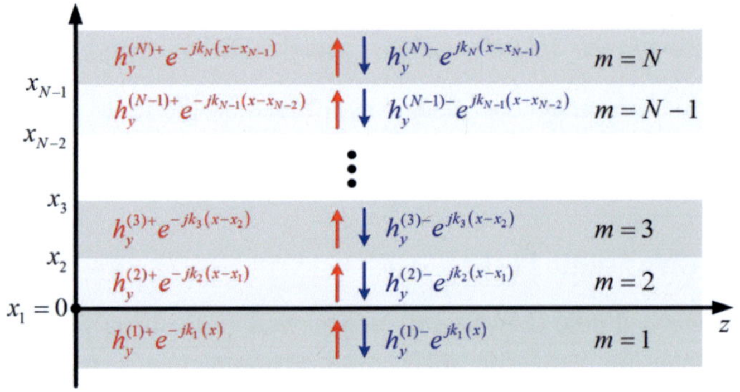

Figure A.1 Multilayer slab waveguide.Layers $m = 1$ and $m = N$ extend to minus and plus infinity, respectively. Layers $m = 2\ldots(N\text{-}1)$ have a finite thicknesses of $x_{\mathrm{m}} - x_{\mathrm{m-1}}$. The solution of the scalar wave equation in each layer is given in the form

$$\breve{h}_y^{m}(x,\omega)=\breve{h}_y^{(m)+}(\omega)e^{-\mathrm{j}k_m(x-x_{m-1})}+\breve{h}_y^{(m)-}(\omega)e^{\mathrm{j}k_m(x-x_{m-1})}, \tag{A.36}$$

where k_m is the wave vector along the x-axis, and $\breve{h}_y^{(m)+}(\omega)$ and $\breve{h}_y^{(m)-}(\omega)$ are the amplitudes of magnetic fields propagating along the positive and negative x-directions, respectively. Substituting the ansatz Eq. (A.36) into the wave equation Eq. (A.35) the modulus of the transverse wave vector in each layer can be written as $k_m=\sqrt{k_0^2\varepsilon_m-\beta^2}$. The amplitudes of electric and magnetic fields in each layer are connected to the amplitudes in the neighbouring layers through boundary conditions: Magnetic and electric field components parallel to the boundary must be continuous. We then express the fields in the $(m+1)^{\text{th}}$ layer in terms of the fields in the mth layer [6]

$$\begin{bmatrix} \breve{h}_y^{(m+1)+} \\ H_y^{(m+1)-} \end{bmatrix} = \frac{1}{2}\begin{bmatrix} 1 & n_{m+1}^2/k_{m+1} \\ 1 & -n_{m+1}^2/k_{m+1} \end{bmatrix}\begin{bmatrix} 1 & 1 \\ k_m/n_m^2 & -k_m/n_m^2 \end{bmatrix} \times\begin{bmatrix} \exp(-\mathrm{j}k_m(x_m - x_{m-1})) & 0 \\ 0 & \exp(\mathrm{j}k_m(x_m - x_{m-1})) \end{bmatrix}\begin{bmatrix} \breve{h}_y^{(m)+} \\ \breve{h}_y^{(m)-} \end{bmatrix} \tag{A.37}$$

As a result, by multiplying the layers' matrices, the amplitudes of the magnetic field within the Nth layer can be expressed by the amplitudes of the field in $m=1$ layer with the help of the **M**-matrix,

$$\begin{bmatrix} \breve{h}_y^{(N)+} \\ \breve{h}_y^{(N)-} \end{bmatrix} = \mathbf{M}\begin{bmatrix} \breve{h}_y^{(1)+} \\ \breve{h}_y^{(1)-} \end{bmatrix} = \begin{bmatrix} M_{11} & M_{12} \\ M_{21} & M_{22} \end{bmatrix}\begin{bmatrix} \breve{h}_y^{(1)+} \\ \breve{h}_y^{(1)-} \end{bmatrix} \tag{A.38}$$

Guided modes have disappearing magnetic fields at infinity, so that the incoming wave amplitudes must vanish, $\breve{h}_y^{(1)+} = 0$ and $\breve{h}_y^{(N)-} = 0$ [173]. These conditions are fulfilled for $M_{22} = 0$ only. The dispersion relation is determined by numerically finding $\beta(\omega)$ for $M_{22} = 0$. The typical dispersion relation of the metal-insulator-metal structure is given in Figure 2.6 of Section 2.2.2.

A.10. Material Parameters and Modelling

In Section 2.1.2, we have discussed the classical the Drude-Lorentz model which can be used to model the wavelength / frequency dependence of optical properties media. However, depending on the frequency range of interest, often it is sufficient to use the Drude model only or neglect the frequency dependence of the refractive index at all. Below we give the material parameters and the approximations used in this thesis.

Dielectrics and semiconductors

In most cases, we neglect the frequency dependence of the optical properties of dielectrics and semiconductors, because of the fact that these materials show resonance frequencies which are far above the frequency range of interest 150…300 THz. Nevethelless, the assumption of the frequency independence is violated in Chapter 2.2 when plotting the dispersion relation of SPP in metal / insulator and metal-insulator-metal slab waveguides, because of the considered vast frequency range. However, the discussion made in Chapter 2.2 is for illustrative purposes only. Simplifying the problem with the assumption of frequency independent optical properties of insulator helps to better understand and interpret the dispersion relation of SPP.

Table A.1 Refractive indices of few dielectric materials at the wavelength of 1.55 μm

	Refractive index n	**Extinction coeff. k**
Silicon (Si)	3.48	0
Silicon dioxide (Si)	1.44	0
Silicon nitride (Si_3N_4)	2	0
EO material M3	1.68	0

Drude model

In the infrared spectral region, the contribution of the bound electrons is included in the frequency independent parameter ε_∞. The plasma frequencies ω_p needed in the Drude model are given in Table A.2. For silver, the plasma frequency is 9.12 eV ($\lambda_p \sim 136$ nm), and for gold 8.6 eV ($\lambda_p \sim 144$ nm) [86]. For comparison, the Drude model parameters for an 8 nm thick ITO layer under a static electric field of 3.3×10^8 V/m are also listed.

Different values have been reported for the plasma frequency of ITO. They are mostly derived by observing SPP resonances in the near - infrared region. The plasma frequency values used in this paper are taken from Michelotti *et al.* [168]. They have investigated SPP resonances for several commercially available ITO samples with various thicknesses. Using these data they have calculated and reported on plasma and collision frequencies of ITO by fitting experimentally obtained transmittance and reflectance spectra with the ones calculated through the Drude model. Observing the plasmon resonances for ITO films with various thicknesses in the wavelength range from 1.4 μm to 2.0 μm, Rhodes *et al.* [174] have reported significantly larger plasma frequencies. Moreover, they have found that the plasma frequency is constant for samples with thicknesses down to 30 nm.

Table A.2 The Drude parameters of ITO, Ag and Au

Parameters	ITO_{off}	ITO_{On} $U = 10$ V, 30 nm(SiO_2)	Ag	Au
Background permittivity ε_∞	3.9	3.9	1	1
Plasma frequency ω_p [s^{-1}]	2.9×10^{15} [25]	2.91×10^{15}	1.38×10^{16}	1.31×10^{16}
Collision frequency γ [s^{-1}]	1.8×10^{14} [25]	1.8×10^{14}	3.22×10^{13}	1.07×10^{14}
Electron effective mass m_{eff}	$0.35\times m_e$ [24]	$0.35\times m_e$	$1\times m_e$	$1\times m_e$
Carrier density N_0 [m^{-3}]	9.25×10^{26}	$\langle N(y)\rangle_{h=8nm} = 9.34\times10^{26}$	6.03×10^{28}	5.39×10^{28}
Complex dielectric permittivity	$-1.66+0.82i$	$-1.72+0.832i$	$-128+3.44i$	$-114+10.1i$

On the other hand, Dasgupta *et al.* [169] have investigated ITO ultrathin films, and the absolute values of plasma frequencies reported by them is approximately two times smaller than the ones shown in Table A.2. However, based on these findings and other experimental investigations [169, 175], one may conclude that for ITO layers thicker than 8 nm the free carrier density of ITO saturates towards its bulk value. Thus, the plasma frequency of ITO used in this paper is the one from Ref. [168] which is near the average of the values so far reported in the literature. It should be emphasized that since the ITO thicknesses of interest are in the range of 8 nm to 15 nm, it is possible to optimize the free carrier density by varying the thickness of ITO such that bound SPP strongly confined in ITO film exists at the wavelength of 1.55 μm. Moreover, the use of the parameters given above is justified by the fact that the key material parameter influencing the overall modulator performance is mainly the plasma frequency of ITO which can be optimized by controlling the ITO sputtering process via the oxygen content, and by the tin-to-indium ratio as well as the temperature and pressure in the deposition chamber.

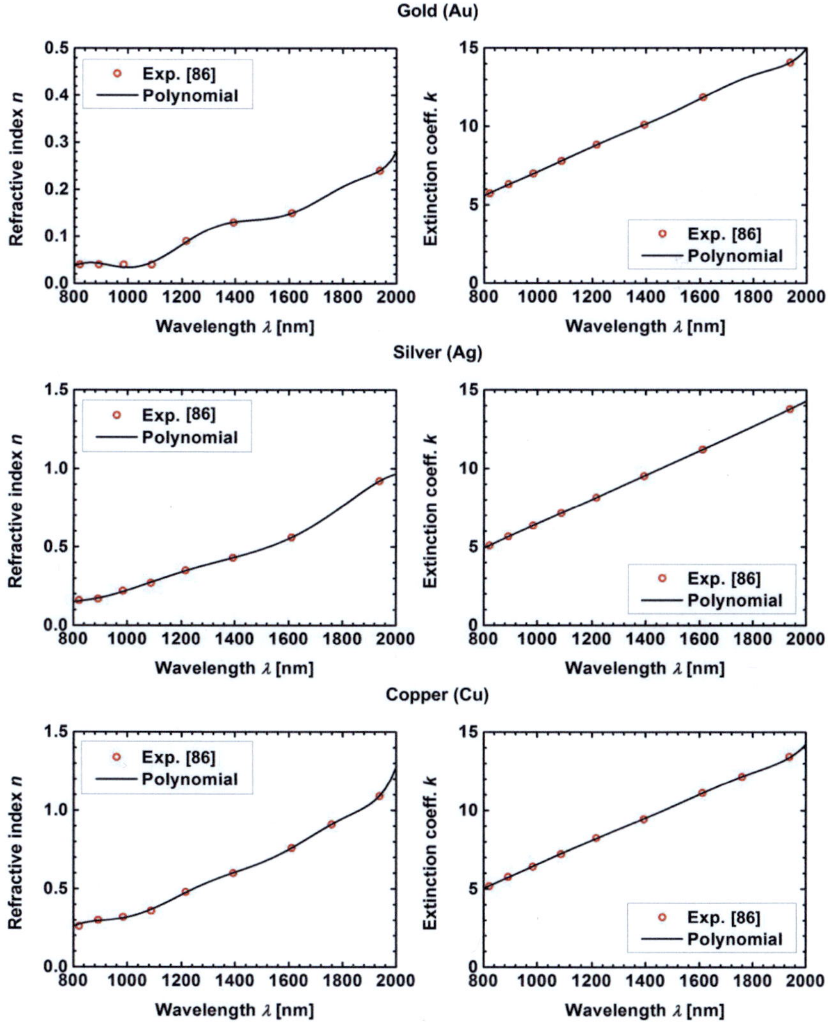

Figure A.2 Measured and fitted complex refractive indices of silver, gold and copper.Red dots represent the measured refractive indices [86], while the black solid lines are the fitted curves with the polynomial of 8th order.

Polynomial fit

The dispersion relation of SPP in the wavelength range of 1 μm to 1.9 μm can more accurately be calculated by modelling the measured optical constants of noble metals with a high order polynomial [176]. We fit the refractive index and the extinction coefficient reported in Ref [86] with an 8th order polynomial. Figure A.2 gives measured and fitted refractive indices and the extinction coefficients of silver, gold and copper.

Appendix B. Fabrication and Experimental Details

B.1. Fabrication of Plasmonic Phase Modulators

Si nanowire waveguides with a height of 220 nm and a width of 450 nm are fabricated on a silicon on insulator (SOI) wafer with a silicon dioxide (SiO_2) thickness of 2 μm. A 193 nm DUV lithography is used followed by Si dry etching. Silicon tapers with a taper angle of 15° and a tip size of 120 nm are structured in the final etching process, see Figure B.1 [101]. The metallic slot waveguides are fabricated by a standard lift-off process with 400 nm thick Poly(methyl methacrylate) (PMMA) e-beam resist. A 150 nm thick gold layer is evaporated onto the samples by an electron beam evaporation system. After completing the lift-off process the samples are coated with the commercially available electro-optic material M3 having a maximum electro-optic coefficient of $r_{33} = 70$ pm/V [177]. The electro-optic property of M3 is then activated by a poling procedure, where a static electric field aligns the randomly oriented dipole moments of the chromophores at an elevated temperature [144]. After rapid cooling to room temperature, the ordering of the dipole moments pertains even after removal of the electric field. The entire process flow employed for fabrication of the plasmonic phase modulator is given in Figure B.2.

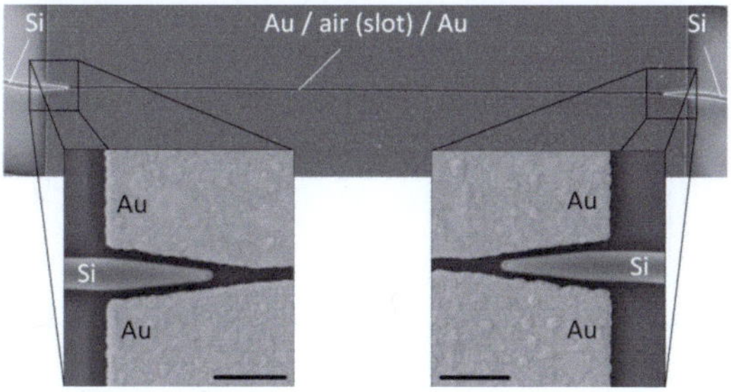

Figure B.1 Scanning electron microscope image of the fabricated plasmonic phase modulator. The photograph is taken before coating the device with an electro-optic material. Black bars are one micrometer.

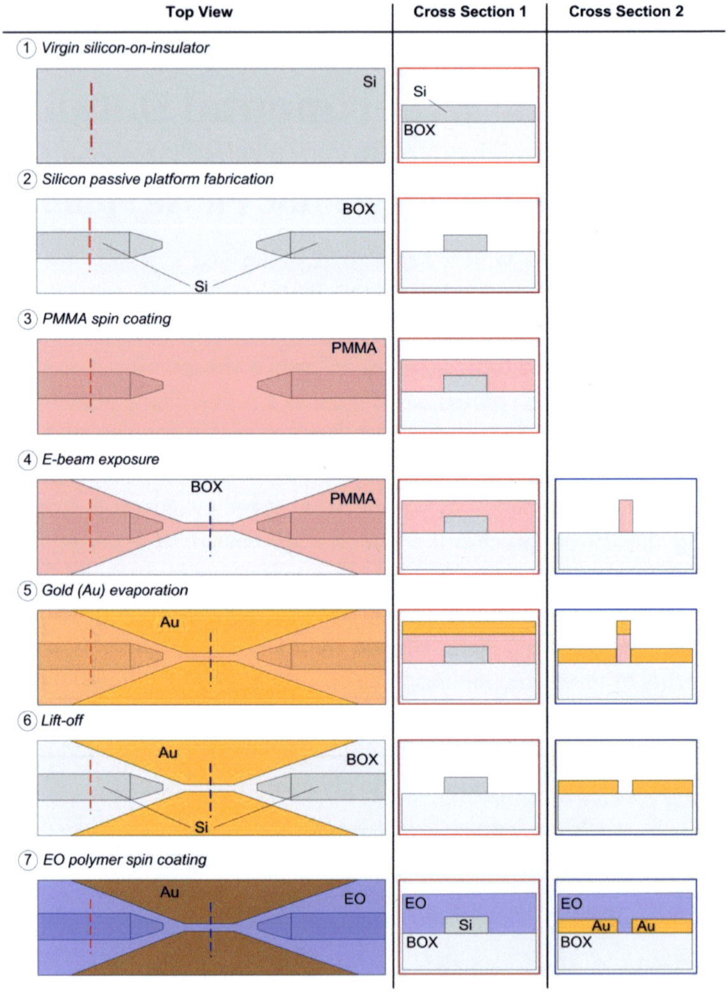

Figure B.2 Fabrication process flow of the plasmonic phase modulator. 1) Silicon-on-insulator wafer with a 220 nm silicon device layer is structure and silicon tapers with various separations are defined (2). (3) The sample is coated with a 400 nm PMMA e-beam resist, followed by an exposure with electron beam energy of 25keV and current of 25pA (4). After the development of the PMMA resist, the sample is coated with 150 nm thick gold layer using an e-beam evaporator (5). After performing the lift-off of the gold layer from the top of the PMMA (6), the sample is coated with an electro-optic (EO) material (7).

B.2. Characterization of Plasmonic Phase Modulators

Optical and RF Characterization

Standard diffraction grating couplers and silicon nanowires are used for coupling the light from the tunable laser source (TLS) into the modulator, and for collecting the modulated light from the phase modulator. The optical spectrum at the output of the chip is recorded with an optical spectrum analyzer (OSA). The transmission spectrum of a plain silicon nanowire without the phase modulator serves as a reference.

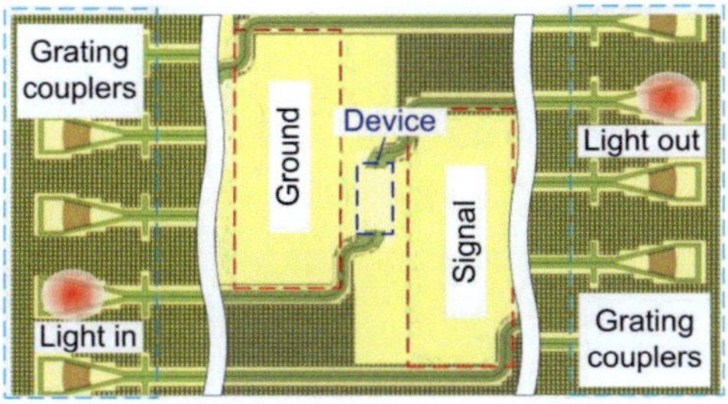

Figure B.3 Optical photograph of the modulator fabricated on the SOI sample. Light is launched into the silicon nanowire using standard diffraction grating couplers. The photonic mode polarized parallel to the substrate is guided to the device. The photonic-plasmonic mode conversion is accomplished by a metal taper coupler. The SPP is then guided through the modulator and converted back to the photonic mode. Light is detected at the output diffraction grating.

We used an impedance standard calibration substrate to define the RF measurement reference plane to the probe tip. The 1stG modulator is contacted with a ground-signal-ground (GSG) RF probe and the complex reflection parameter S_{11} is measured with an Anritsu 37397C Vector Network Analyzer (VNA).

Electro-optical Characterization

The optical carrier with frequency $f_c = \omega_c / (2\pi)$ is phase modulated with a frequency f_m and a phase modulation index $\eta \sim U_m$, resulting in an optical signal $\cos(\omega_c t + \eta \sin \omega_m t)$. Employing Jacobi-Anger expansion of an exponential phasor $e^{j(\omega_c t+\eta\sin\omega_m t)}$, it can be shown that an optical signal with a sinusoidally modulated phase has a spectrum which consists of discrete frequency lines

$$e^{j(\omega_c t+\eta\sin\omega_m t)}=\sum_{n=-\infty}^{+\infty}J_n(\eta)e^{j(\omega_c+n\omega_m)t} \tag{B.1}$$

The spectral amplitude for each of the spectral line is given by the corresponding Bessel function $J_n(\eta)$. A spectrum for an small-signal phase modulated optical signal measured at the output of the plasmonic phase modulator is given in Figure B.4 (a). Here, the carrier has a relative amplitude in proportion to the 0th-order Bessel function $J_0(\eta)$, and the first sideband has an amplitude in proportion to the 1st-order Bessel function $J_1(\eta)$. From the ratio of the respective line heights in the power spectrum, $J_0^2(\eta)/J_1^2(\eta)$, one therefore can extract the modulation index η. A modulation index η versus the power ratio $S(f_c)/S(f_c+f_m)$ is given in Figure B.4 (b), which can be used to estimate the modulation index from the measured spectrum of the optical signal.

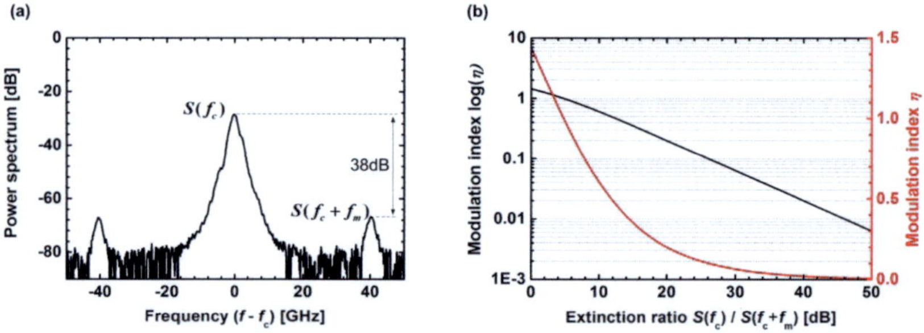

Figure B.4 Spectrum of the sinusoidal phase modulated signal and its relation to the modulation index. (a) Optical spectrum measured after the1stG plasmonic phase modulator driven with an RF power of 8 dBm is given as an example of the spectrum of the sinusoidal phase modulated signal. (b) Modulation index is given as a function of the power ratio between carrier frequency and the first side band $S(f_c)/S(f_c+f_m)$ carrier and first side.

The RF frequency response of the device is studied by drive the modulator with a sinusoidal signal having a frequency in the range $f_m = (1\ldots65)$ GHz and an amplitude of $U_m = 0.1$ V. We have measured the phase modulation index η for each frequency. Due to the capacitive termination, the resulting voltage across the device is doubled when comparing with the voltage across a matched 50 Ohm terminating resistor. This amplitude is kept constant during the RF frequency sweep by calibrating the electrical power absorbed in a matched load before the RF probe was connected.

Data Modulation Experiments

Infrared light with a wavelength of 1550.92 nm is first amplified to a power level of 15dBm and then launched into the chip (measured in the fibre before coupling the signal into the chip). The phase of the surface plasmon polariton (SPP) is encoded with a PRBS signal at the voltage swing of U_{pp} measured across a 50 Ω resistor. 1stG modulators are driven with an electrical signal with a voltage swing of $U_{\text{pp}} = 7.5$V resulting in a peak-to-peak phase modulation of 0.23 rad. The resulting binary phase shift keyed (BPSK) signal is amplified in an EDFA which acts as an optical pre-amplifier and as a part of the receiver constituted by an Agilent N4391A Optical Modulation Analyzer (OMA). It should be noted that the phase modulation index achieved in the generated BPSK signal is less than a π resulting in a signal's large DC in the coherent receiver. This large DC part is removed by using OMA's AC coupling option. In case of the 2ndG modulator, the resulting BPSK signal was subsequently converted into an intensity modulated signal by a delay interferometer (DI) with a free spectral range (FSR) of 40 GHz. The signal is then directly detected with a single photodiode. To compensate the losses of the in-house built DI, an EDFA was used. The eye diagram and the bit error ratio are measured with an Infiniium DCA-J Agilent 86100C digital communication analyzer (DCA) and an Anritsu MP1776A bit error ratio tester (BERT), respectively.

Glossary

Calligraphic Symbols

$\mathcal{E}$	Low-frequency modulating electric field

Latin Symbols

A	Integration area
$\mathbf{B}$	Magnetic inductance
c	Speed of light in vacuum
C	Capacitance
d	Distance between silicon and plasmonic waveguides
$\mathbf{D}$	Displacement field
d_s	Distance between silicon taper tip and metallic slot waveguide
e	Electron charge
$\mathbf{E}$	Electric field
$\breve{\mathbf{e}}_{\mathrm{rad}}$	Electric field profile of radiating waves
$\breve{\mathbf{e}}_{\mu}$	Electric field profile of a μth eigenmodes
f	Frequency
h	Metallic slot size
$\mathbf{H}$	Magnetic field
h_c	Thin conductor thickness
$\breve{\mathbf{h}}_{\mathrm{rad}}$	Magnetic field profile of radiating waves
$\breve{\mathbf{h}}_{\mu}$	Magnetic field profile of a μth eigenmode
$\hbar$	Planck's constant
$\mathbf{J}$	Electric current

k_0	Vacuum wave vector
k_i	Transverse wave vector in an ith layer
l	Metal width in a metallic slot waveguide
L	Length of a device
$L_{1\text{dB}}$	Length of a device resulting in an extinction ratio of 1 dB
L_e	Propagation length
m_e	Electron mass
m_{eff}	Electron effective mass
$\underline{n}$	Complex refractive index tensor
n	Complex refractive index
$\underline{n}'$	Real part of the refractive index tensor
n'	Real part of the refractive index of an isotropic material
$\underline{n}''$	Imaginary part of the refractive index tensor
n''	Imaginary part of the refractive index of an isotropic material
n_{cl}	Refractive index of a cladding
n_{cr}	Refractive index of a core
n_{eff}	Effective refractive index
n_{EO}	Refractive index of an electro-optic material
n_g	Group refractive index
$\underline{n}_{\text{NL}}$	Nonlinear refractive index
n_{rl}	Refractive index of a core
n_{sb}	Refractive index of a substrate
N	Free electron density in a conductor
N_i	Density of Lorentz oscillators with a resonance frequency ω_{L_i}
ΔN	Change of the free electron density

$\mathbf{p}$	Dipole moment
$\mathbf{P}$	Polarization
$\mathbf{P}_{\mathrm{L}}$	Linear term of the polarization
$\mathbf{P}_{\mathrm{NL}}$	Nonlinear term of the polarization
$\mathbf{r}$	Position vector
R	Resistance
r_{ijk}, r_{hk}	Electro-optic tensor, electro-optic matrix
$\mathbf{S}$	Poynting vector
t	Time
T	Temperature
w	Metal thicknesses in a metallic slot waveguide
w_{Si}	Width of a silicon strip waveguide
x, y, z	Cartesian coordinates

Greek Symbols

α_{dB}	Attenuation constant in dB
α_{off}	Attenuation constant in voltage-off state
α_{on}	Attenuation constant in voltage-on state
$\Delta\alpha_{10\mathrm{V}}$	Change of a constant for an applied voltage of 10V
β_{μ}	Propagation constant of a μth eigenmodes
γ	Electron damping constant
Γ	Interaction factor
$\underline{\varepsilon}$	Relative permittivity tensor
ε	Relative permittivity of an isotropic material
$\underline{\varepsilon}', \underline{\varepsilon}''$	Real and imaginary parts of the relative permittivity tensor

$\varepsilon', \varepsilon''$	Real and imaginary parts of the relative permittivity of an isotropic material
ε_0	Vacuum permittivity
$\varepsilon_{\text{stat}}$	Static permittivity
θ	Tapering angle of a metal taper mode converter
λ	Vacuum wavelength of light
λ_c	Carrier wavelength
μ_0	Vacuum permeability
ρ_{net}	Electric current
σ	Electrical conductivity
ϕ	Electric potential
$\chi^{(n)}$	n^{th} order influence function
$\underline{\breve{\chi}}^{(n)}$	n^{th} order susceptibility tensor
ω	Angular frequency
ω_{L_i}	Resonance frequency of a Lorenz oscillator
ω_p	Plasma frequency
ω_{sp}	Surface plasmon frequency

Acronyms

1stG	First generation
2nG	Second generation
BER	Bit error rate
BPSK	Binary phase shift keying
CMOS	Complementary metal oxide semiconductor
CPU	Central processing unit
DI	Delay interferometer
DPSK	Differential phase shift keying
DUT	Device under test
DUV	Deep ultraviolet
EAM	Electro-absorption modulator
EM	Electromagnetic
EO	Electro-optic
EOM	Electro-optic modulator
EVM	Error vector magnitudes
FDTD	Finite difference time domain
FEM	Finite element method
FIT	Finite integration technique
FK	Franz-Keldysh effect
FSR	Free spectral range
IC	Integrated circuit
IM	Optical intensity modulator
IMI	Insulator-metal-insulator
IO	Input output

IQ	In-phase and quadrature-phase
ITO	Indium tin oxide
SP	Surface plasmon
SPP	Surface plasmon polariton
MIM	Metal-insulator-metal
MSW	Metallic slot waveguide
MTMC	Metallic tapered mode converter
MWS	Microwave Studio
MZI	Mach-Zehnder interferometer
MZM	Mach-Zehnder modulator
OSA	Optical spectrum analyser
PBS	Polarization beam splitter
PC	Photonic crystal
PM	Optical phase modulator
PMMA	Poly(methyl methacrylate)
PPM	Plasmonic phase modulator
PRBS	Pseudo-random bit sequence
QAM	Quadrature amplitude
RF	Radio frequency
RRM	Ring resonator modulator
SEM	Scanning electron microscope
SOH	Silicon organic hybrid
SPPAM	Surface plasmon polariton absorption modulator
SWG	Silicon strip waveguide
TE	Transverse electric
TLS	Tuneable light source
TM	Transverse magnetic

References

[1] S. Maier, Plasmonics: fundamentals and applications. New York: Springer, 2007.

[2] H. Raether, Surface plasmons on smooth and rough surfaces and on gratings. Berlin: Springer, 1988.

[3] G. E. Moore, "Cramming more components onto integrated circuits," Electronics, vol. 38, no. 8, pp. 114–117, 1965.

[4] R. H. Dennard, F. H. Gaensslen, H.-N. Yu, V. L. Rideout, E. Bassous, and A. R. Leblanc, "Design of ion-implanted MOSFET'S with very small physical dimensions," IEEE J. Solid-State Circuits, vol. 9, no. 5, pp. 256–268, 1974.

[5] M. Bohr, "A 30 year retrospective on dennard's MOSFET scaling paper," IEEE Solid-State Circuits Soc. Newsl., vol. 12, no. 1, pp. 11–13, 2007.

[6] Y. Taur, D. A. Buchanan, W. E. I. Chen, D. J. Frank, K. E. Ismail, S. Lo, G. A. Sai-halasz, R. G. Viswanathan, H. C. Wann, S. J. Wind, and H. Wong, "CMOS scaling into the nanometer regime," Proc. IEEE, vol. 85, no. 4, pp. 486–504, 1997.

[7] P. Gepner and M. F. Kowalik, "Multi-core processors: new way to achieve high system performance," in Int. Symp. Parallel Comput. Electr. Eng., pp. 9–13, 2006.

[8] E. Mohammed, A. Alduino, T. Thomas, H. Braunisch, D. Lu, J. Heck, A. Liu, I. Young, B. Barnett, G. Vandentop, and R. Mooney, "Optical interconnect system integration for ultra-short-reach applications," Intel Technol. J., vol. 8, no. 2, pp. 115–127, 2004.

[9] D. Hillerkuss, R. Schmogrow, T. Schellinger, M. Jordan, M. Winter, G. Huber, T. Vallaitis, R. Bonk, P. Kleinow, F. Frey, M. Roeger, S. Koenig, A. Ludwig, A. Marculescu, J. Li, M. Hoh, M. Dreschmann,

J. Meyer, S. Ben Ezra, N. Narkiss, B. Nebendahl, F. Parmigiani, P. Petropoulos, B. Resan, A. Oehler, K. Weingarten, T. Ellermeyer, J. Lutz, M. Moeller, M. Huebner, J. Becker, C. Koos, W. Freude, and J. Leuthold, "26 Tbit/s line-rate super-channel transmission utilizing all-optical fast Fourier transform processing," Nat. Photonics, vol. 5, no. 6, pp. 364–371, 2011.

[10] I. A. Young, E. Mohammed, J. T. S. Liao, A. M. Kern, S. Palermo, B. A. Block, M. R. Reshotko, and P. L. D. Chang, "Optical I/O technology for tera-scale computing," IEEE J. Solid-State Circuits, vol. 45, no. 1, pp. 235–248, Jan. 2010.

[11] G. T. Reed, G. Mashanovich, F. Y. Gardes, and D. J. Thomson, "Silicon optical modulators," Nat. Photonics, vol. 4, no. 8, pp. 518–526, Jul. 2010.

[12] J. Van Campenhout, W. M. J. Green, S. Assefa, and Y. a Vlasov, "Integrated NiSi waveguide heaters for CMOS-compatible silicon thermo-optic devices.," Opt. Lett., vol. 35, no. 7, pp. 1013–5, Apr. 2010.

[13] A. Masood, M. Pantouvaki, D. Goossens, G. Lepage, P. Verheyen, and P. Absil, "CMOS compatible tungsten heaters for silicon photonic waveguides," in IEEE 9th Int. Conf. Gr. IV Photonics, pp. 234–236, 2012.

[14] P. Dong, W. Qian, H. Liang, R. Shafiiha, D. Feng, G. Li, J. E. Cunningham, A. V Krishnamoorthy, and M. Asghari, "Thermally tunable silicon racetrack resonators with ultralow tuning power," Opt. Express, vol. 18, no. 19, pp. 20298–304, Sep. 2010.

[15] M. S. Rasras, D. M. Gill, M. P. Earnshaw, C. R. Doerr, J. S. Weiner, C. a. Bolle, and Y.-K. Chen, "CMOS silicon receiver integrated with Ge detector and reconfigurable optical filter," IEEE Photonics Technol. Lett., vol. 22, no. 2, pp. 112–114, Jan. 2010.

[16] T. Nikolajsen, K. Leosson, and S. I. Bozhevolnyi, "Surface plasmon polariton based modulators and switches operating at telecom wavelengths," Appl. Phys. Lett., vol. 85, no. 24, pp. 5833–5835, 2004.

[17] A. L. Lereu, A. Passian, J.-P. Goudonnet, T. Thundat, and T. L. Ferrell, "Optical modulation processes in thin films based on thermal effects of surface plasmons," Appl. Phys. Lett., vol. 86, no. 15, p. 154101, 2005.

[18] L. Alloatti, D. Korn, R. Palmer, D. Hillerkuss, J. Li, A. Barklund, R. Dinu, M. Fournier, J. Fedeli, H. Yu, W. Bogaerts, P. Dumon, and R. Baets, "42.7 Gbit/s electro-optic modulator in silicon technology," Opt. Express, vol. 19, no. 12, pp. 11841–11851, 2011.

[19] J. M. Brosi, C. Koos, L. C. Andreani, M. Waldow, J. Leuthold, and F. Wolfgang, "High-speed low-voltage electro-optic modulator with a polymer-infiltrated silicon photonic crystal waveguide," Opt. Express, vol. 16, no. 6, pp. 4177–4191, 2008.

[20] T. Baehr-Jones, B. Penkov, J. Huang, P. Sullivan, J. Davies, J. Takayesu, J. Luo, T.-D. Kim, L. Dalton, A. Jen, M. Hochberg, and A. Scherer, "Nonlinear polymer-clad silicon slot waveguide modulator with a half wave voltage of 0.25 V," Appl. Phys. Lett., vol. 92, no. 16, p. 163303, 2008.

[21] R. Ding, T. Baehr-Jones, Y. Liu, R. Bojko, J. Witzens, S. Huang, J. Luo, S. Benight, P. Sullivan, J.-M. Fedeli, M. Fournier, L. Dalton, A. Jen, and M. Hochberg, "Demonstration of a low V pi L modulator with GHz bandwidth based on electro-optic polymer-clad silicon slot waveguides," Opt. Express, vol. 18, no. 15, pp. 15618–15623, Jul. 2010.

[22] D. Korn, R. Palmer, H. Yu, P. C. Schindler, L. Alloatti, M. Baier, R. Schmogrow, W. Bogaerts, S. K. Selvaraja, G. Lepage, M. Pantouvaki, J. M. D. Wouters, P. Verheyen, J. Van Campenhout, B. Chen, R. Baets, P. Absil, R. Dinu, C. Koos, W. Freude, and J. Leuthold, "Silicon-organic hybrid (SOH) IQ modulator using the linear electro-optic effect for transmitting 16QAM at 112Gbit/s," Opt. Express, vol. 21, no. 11, pp. 13219–13227, 2013.

[23] R. Palmer, L. Alloatti, D. Korn, P. C. Schindler, M. Baier, J. Bolten, T. Wahlbrink, M. Waldow, R. Dinu, W. Freude, C. Koos, and J. Leuthold, "Low power Mach–Zehnder modulator in silicon-organic hybrid tech-

nology," IEEE Photonics Technol. Lett., vol. 25, no. 13, pp. 1226–1229, Jul. 2013.

[24] W. Cai, J. S. White, and M. L. Brongersma, "Compact, high-speed and power-efficient electrooptic plasmonic modulators," Nano Lett., vol. 9, no. 12, pp. 4403–4411, 2009.

[25] S. Randhawa, S. Lachèze, J. Renger, A. Bouhelier, R. E. de Lamaestre, A. Dereux, and R. Quidant, "Performance of electro-optical plasmonic ring resonators at telecom wavelengths," Opt. Express, vol. 20, no. 3, pp. 2354–2362, Jan. 2012.

[26] S.-I. Inoue and S. Yokoyama, "Numerical simulation of ultra-compact electro-optic modulator based on nanoscale plasmon metal gap waveguides," Electron. Lett., vol. 45, no. 21, pp. 1087–1089, 2009.

[27] Q. Xu, B. Schmidt, S. Pradhan, and M. Lipson, "Micrometre-scale silicon electro-optic modulator," Nature, vol. 435, no. 7040, pp. 325–327, May 2005.

[28] Q. Xu, S. Manipatruni, and B. Schmidt, "12.5 Gbit/s carrier-injection-based silicon micro-ring silicon modulators," Opt. Express, vol. 15, no. 2, pp. 430–436, 2007.

[29] S. Manipatruni, Q. Xu, B. Schmidt, J. Shakya, and M. Lipson, "High Speed Carrier Injection 18 Gb/s Silicon Micro-ring Electro-optic Modulator," LEOS 2007 - IEEE Lasers Electro-Optics Soc. Annu. Meet. Conf. Proc., pp. 537–538, Oct. 2007.

[30] M. R. Watts, D. C. Trotter, R. W. Young, and A. L. Lentine, "Ultralow power silicon microdisk modulators and switches," in IEEE 5th Int. Conf. Gr. IV Photonics, p. WA2, 2008.

[31] P. Dong, S. Liao, D. Feng, H. Liang, D. Zheng, R. Shafiiha, C.-C. Kung, W. Qian, G. Li, X. Zheng, A. V. Krishnamoorthy, and M. Asghari, "Low V pp, ultralow-energy, compact, high-speed silicon electro-optic modulator," Opt. Express, vol. 17, no. 25, pp. 22484–22490, 2009.

[32] A. Liu, L. Liao, D. Rubin, H. Nguyen, B. Ciftcioglu, Y. Chetrit, N. Izhaky, and M. Paniccia, "High-speed optical modulation based on carrier depletion in a silicon waveguide," Opt. Express, vol. 15, no. 2, pp. 660–668, Jan. 2007.

[33] L. Liao, A. Liu, J. Basak, H. Nguyen, R. Cohen, N. Izhaky, and M. Paniccia, "40 Gbit s-1 silicon optical modulator for highspeed applications," Electron. Lett., vol. 43, no. 22, 2007.

[34] W. M. J. Green, M. J. Rooks, L. Sekaric, and Y. A. Vlasov, "Ultra-compact, low RF power, 10 Gb/s silicon Mach-Zehnder modulator," Opt. Express, vol. 15, no. 25, pp. 17106–17113, Dec. 2007.

[35] D. M. Gill, M. Rasras, K.-Y. Tu, Y.-K. Chen, A. E. White, S. S. Patel, D. Carothers, A. Pomerene, R. Kamocsai, C. Hill, and J. Beattie, "Internal bandwidth equalization in a CMOS-compatible Si-ring modulator," IEEE Photonics Technol. Lett., vol. 21, no. 4, pp. 200–202, 2009.

[36] P. Dong, L. Chen, and Y. Chen, "High-speed low-voltage single-drive push-pull silicon Mach-Zehnder modulators," Opt. Express, vol. 20, no. 6, pp. 6163–6169, 2012.

[37] J. A. Dionne, K. Diest, L. A. Sweatlock, and H. A. Atwater, "PlasMOStor: A metal–oxide–Si field effect plasmonic modulator," Nano Lett., vol. 9, no. 2, pp. 897–902, 2009.

[38] S. Zhu, G. Q. Lo, and D. L. Kwong, "Theoretical investigation of silicon MOS-type plasmonic slot waveguide based MZI modulators," Opt. Express, vol. 18, no. 26, pp. 27802–27819, Dec. 2010.

[39] S. Zhu, G. Q. Lo, and D. L. Kwong, "Phase modulation in horizontal metal-insulator- silicon-insulator-metal plasmonic waveguides," Opt. Express, vol. 21, no. 7, pp. 8320–8330, 2013.

[40] T. Baba, S. Akiyama, M. Imai, N. Hirayama, H. Takahashi, Y. Noguchi, T. Horikawa, and T. Usuki, "50-Gb/s ring-resonator-based silicon modulator," Opt. Express, vol. 21, no. 10, pp. 11869–11876, 2013.

[41] D. J. Thomson, F. Y. Gardes, D. C. Cox, J.-M. Fedeli, G. Z. Mashanovich, and G. T. Reed, "Self-aligned silicon ring resonator optical modulator with focused ion beam error correction," J. Opt. Soc. Am. B, vol. 30, no. 2, p. 445, Jan. 2013.

[42] X. Xiao, X. Li, H. Xu, Y. Hu, K. Xiong, Z. Li, T. Chu, J. Yu, and Y. Yu, "44-Gb/s silicon microring modulators based on zigzag pn junctions," IEEE Photonics Technol. Lett., vol. 24, no. 19, pp. 1712–1714, Oct. 2012.

[43] D. Marris-Morini, C. Baudot, J.-M. Fédéli, G. Rasigade, N. Vulliet, A. Souhaité, M. Ziebell, P. Rivallin, S. Olivier, P. Crozat, X. Le Roux, D. Bouville, S. Menezo, F. Bœuf, and L. Vivien, "Low loss 40 Gbit/s silicon modulator based on interleaved junctions and fabricated on 300 mm SOI wafers," Opt. Express, vol. 21, no. 19, pp. 22471–22475, 2013.

[44] H. C. Nguyen, S. Hashimoto, M. Shinkawa, and T. Baba, "Compact and fast photonic crystal silicon optical modulators," Opt. Express, vol. 20, no. 20, pp. 22465–22474, Sep. 2012.

[45] X. Xiao, H. Xu, X. Li, Z. Li, T. Chu, Y. Yu, and J. Yu, "High-speed, low-loss silicon Mach–Zehnder modulators with doping optimization," Opt. Express, vol. 21, no. 4, pp. 4116–4125, Feb. 2013.

[46] D. Feng, W. Qian, H. Liang, C.-C. Kung, Z. Zhou, Z. Li, J. S. Levy, R. Shafiiha, J. Fong, B. Jonathan Luff, and M. Asghari, "High-speed GeSi electroabsorption modulator on the SOI waveguide platform," IEEE J. Sel. Top. Quantum Electron., vol. 19, no. 6, pp. 3401710–3401710, Nov. 2013.

[47] Y.-H. Kuo, Y. K. Lee, Y. Ge, S. Ren, J. E. Roth, T. I. Kamins, D. a B. Miller, and J. S. Harris, "Strong quantum-confined Stark effect in germanium quantum-well structures on silicon," Nature, vol. 437, no. 7063, pp. 1334–1336, Oct. 2005.

[48] Y. Li and B. Cheng, "Design of electro-absorption modulator with tapered-mode coupler on the GeSi layer," J. Opt., vol. 15, no. 8, p. 085501, Aug. 2013.

[49] J. Liu, M. Beals, A. Pomerene, S. Bernardis, R. Sun, J. Cheng, L. C. Kimerling, and J. Michel, "Waveguide-integrated, ultralow-energy GeSi electro-absorption modulators," Nat. Photonics, vol. 2, no. 7, pp. 433–437, May 2008.

[50] J. Liu, D. Pan, S. Jongthammanurak, K. Wada, L. C. Kimerling, and J. Michel, "Design of monolithically integrated GeSi electro-absorption modulators and photodetectors on a SOI platform," Opt. Express, vol. 15, no. 2, pp. 623–628, Jan. 2007.

[51] M. Schmid, M. Kaschel, M. Gollhofer, M. Oehme, J. Werner, E. Kasper, and J. Schulze, "Franz–Keldysh effect of germanium-on-silicon p–i–n diodes within a wide temperature range," Thin Solid Films, vol. 525, pp. 110–114, Dec. 2012.

[52] G. Si, M. S. Rouifed, P. Chaisakul, D. Marris-morini, J. Frigerio, G. Isella, D. Chrastina, S. Edmond, X. Le Roux, J. Coudevylle, and L. Vivien, "Quantum-confined Stark effect at 1.3μm in Ge/Si0.35Ge0.65 quantum-well structure," Opt. Lett., vol. 37, no. 19, pp. 3960–3962, 2012.

[53] N.-N. Feng, D. Feng, S. Liao, X. Wang, P. Dong, H. Liang, C.-C. Kung, W. Qian, J. Fong, R. Shafiiha, Y. Luo, J. Cunningham, A. V Krishnamoorthy, and M. Asghari, "30GHz Ge electro-absorption modulator integrated with 3μm silicon-on-insulator waveguide," Opt. Express, vol. 19, no. 8, pp. 7062–7067, Apr. 2011.

[54] E. Feigenbaum, K. Diest, and H. A. Atwater, "Unity-order index change in transparent conducting oxides at visible frequencies," Nano Lett., vol. 10, no. 6, pp. 2111–2116, Jun. 2010.

[55] V. E. Babicheva and A. V. Lavrinenko, "Plasmonic modulator optimized by patterning of active layer and tuning permittivity," Opt. Commun., vol. 285, no. 24, pp. 5500–5507, Nov. 2012.

[56] F. Yi, E. Shim, A. Y. Zhu, H. Zhu, J. C. Reed, and E. Cubukcu, "Voltage tuning of plasmonic absorbers by indium tin oxide," Appl. Phys. Lett., vol. 102, no. 22, p. 221102, 2013.

[57] M. Liu, X. Yin, E. Ulin-Avila, B. Geng, T. Zentgraf, L. Ju, F. Wang, and X. Zhang, "A graphene-based broadband optical modulator," Nature, vol. 474, no. 7349, pp. 64–67, Jun. 2011.

[58] A. V Krasavin, a V Zayats, and N. I. Zheludev, "Active control of surface plasmon–polariton waves," J. Opt. A Pure Appl. Opt., vol. 7, no. 2, pp. S85–S89, Feb. 2005.

[59] K. J. A. Ooi, P. Bai, H. S. Chu, and L. K. Ang, "Ultracompact vanadium dioxide dual-mode plasmonic waveguide electroabsorption modulator," Nanophotonics, vol. 2, no. 1, pp. 13–19, Jan. 2013.

[60] L. A. Sweatlock and K. Diest, "Vanadium dioxide based plasmonic modulators," Opt. Express, vol. 20, no. 8, pp. 8700–8709, Apr. 2012.

[61] A. Pashkin, C. Kübler, H. Ehrke, R. Lopez, A. Halabica, R. F. Haglund, R. Huber, and A. Leitenstorfer, "Ultrafast insulator-metal phase transition in VO2 studied by multiterahertz spectroscopy," Phys. Rev. B, vol. 83, no. 19, p. 195120, May 2011.

[62] W. De Cort, J. Beeckman, T. Claes, K. Neyts, and R. Baets, "Wide tuning of silicon-on-insulator ring resonators with a liquid crystal cladding," Opt. Lett., vol. 36, no. 19, pp. 3876–3878, Oct. 2011.

[63] B. Maune, R. Lawson, C. Gunn, A. Scherer, and L. Dalton, "Electrically tunable ring resonators incorporating nematic liquid crystals as cladding layers," Appl. Phys. Lett., vol. 83, no. 23, p. 4689, 2003.

[64] J. Pfeifle, L. Alloatti, W. Freude, J. Leuthold, and C. Koos, "Silicon-organic hybrid phase shifter based on a slot waveguide with a liquid-crystal cladding," Opt. Express, vol. 20, no. 14, pp. 15359–15376, Jul. 2012.

[65] W. Franz, "Einfluß eines elektrischen Feldes auf eine optische Absorptionskante," Z. Naturforsch., vol. 13, pp. 484–489, 1958.

[66] L. V. Keldysh, "Behaviour of non-metallic crystals in strong electric fields," Sov. J. Exp. Theor. Phys., vol. 6, p. 763, 1958.

[67] B. E. A. Saleh and M. C. Teich, Fundamentals of photonics. John Wiley & Sons, 1991.

[68] V. J. Sorger, N. D. Lanzillotti-Kimura, R.-M. Ma, and X. Zhang, "Ultra-compact silicon nanophotonic modulator with broadband response," J. Nanophotonics, vol. 1, no. 1, pp. 17–22, Jan. 2012.

[69] R. F. Oulton, V. J. Sorger, D. A. Genov, D. F. P. Pile, and X. Zhang, "A hybrid plasmonic waveguide for subwavelength confinement and long-range propagation," Nat. Photonics, vol. 2, no. 8, pp. 496–500, Jul. 2008.

[70] W. Freude, "Optical transmitters and receivers," Lect. Notes, Karlsruhe Insitute Technol., 2013.

[71] L. N. Binh, Digital processing: optical transmission and receiving techniques. CRC Press, 2014.

[72] D. J. Thomson, F. Y. Gardes, Y. Hu, G. Mashanovich, M. Fournier, P. Grosse, J.-M. Fedeli, and G. T. Reed, "High contrast 40Gbit/s optical modulation in silicon," Opt. Express, vol. 19, no. 12, pp. 11507–11516, 2011.

[73] J. Wang, H. Li, L. Li, A. Pang, Z. Sbeng, S. Zou, and F. Gan, "High-speed silicon Mach-Zehnder modulator operating beyond 40Gb/s." p. ATh2B.3, 2012.

[74] F. Y. Gardes, D. J. Thomson, N. G. Emerson, and G. T. Reed, "40Gb/s silicon photonics modulator for TE and TM polarisations.," Opt. Express, vol. 19, no. 12, pp. 11804–11814, Jun. 2011.

[75] M. Ziebell, D. Marris-Morini, G. Rasigade, J.-M. Fédéli, P. Crozat, E. Cassan, D. Bouville, and L. Vivien, "40 Gbit/s low-loss silicon optical modulator based on a pipin diode.," Opt. Express, vol. 20, no. 10, pp. 10591–10596, May 2012.

[76] R. Palmer, S. Koeber, W. Heni, D. L. Elder, D. Korn, H. Yu, L. Alloatti, S. Koenig, P. C. Schindler, W. Bogaerts, M. Pantouvaki, G. Lepage, P. Verheyen, J. Van Campenhout, P. Absil, R. Baets, L. R. Dalton, W. Freude, J. Leuthold, and C. Koos, "High-speed silicon-organic hybrid

(SOH) modulator with 1.6 fJ/bit and 180 pm/V in-device nonlinearity," in Eur. Conf. Opt. Commun., pp. 1–3, 2013.

[77] J. Leuthold, C. Koos, W. Freude, L. Alloatti, R. Palmer, D. Korn, J. Pfeifle, M. Lauermann, R. Dinu, S. Wehrli, M. Jazbinsek, G. Peter, M. Waldow, T. Wahlbrink, J. Bolten, H. Kurz, M. Fournier, J. Fedeli, H. Yu, and W. Bogaerts, "Silicon-organic hybrid electro-optical devices," IEEE J. Sel. Top. Quantum Electron., vol. 19, no. 6, p. 3401413, 2013.

[78] A. Brimont, D. J. Thomson, P. Sanchis, J. Herrera, F. Y. Gardes, J. M. Fedeli, G. T. Reed, and J. Martí, "High speed silicon electro-optical modulators enhanced via slow light propagation," Opt. Express, vol. 19, no. 21, pp. 21986–21991, 2011.

[79] A. Brimont, D. J. Thomson, F. Y. Gardes, J. M. Fedeli, G. T. Reed, J. Martí, and P. Sanchis, "High-contrast 40Gb/s operation of a 500µm long silicon carrier-depletion slow wave modulator," Opt. Lett., vol. 37, no. 17, pp. 3504–3506, Sep. 2012.

[80] R. Palmer, S. Koeber, M. Woessner, D. L. Elder, W. Heni, D. Korn, H. Yu, M. Lauermann, W. Bogaerts, L. R. Dalton, W. Freude, J. Leuthold, and C. Koos, "High-speed silicon-organic hybrid(SOH) modulators with 230pm/V electro-optic coefficient using advanced materials," in Opt. Fiber Commun. Conf., San Francisco, CA, p. M3G.4, 2014.

[81] John David Jackson, Classical electrodynamics, 3rd ed. John Wiley & Sons, 1999.

[82] R. W. Boyd, Nonlinear Optics, 3rd ed. Academic Press, 2008.

[83] Y.-R. Shen, The principles of nonlinear optics. New York: Wiley-Interscience, 1984.

[84] C. F. Klingshirn, Semiconductor optics. Springer, 2007.

[85] F. Wooten, Optical properties of solids. Academic Press, 1972.

[86] P. B. Johnson and R. W. Christy, "Optical constants of the noble metals," Phys. Rev. B, vol. 6, no. 12, pp. 4370–4379, 1972.

[87] L. R. Dalton, P. A. Sullivan, and D. H. Bale, "Electric field poled organic electro-optic materials: state of the art and future prospects," Chem. Rev., vol. 110, no. 1, pp. 25–55, Jan. 2010.

[88] M. Born and E. Wolf, Principles of optics: electromagnetic theory of propagation, interference and diffraction of light. Cambridge University Press, 1999.

[89] F. Zernike and J. E. Midwinter, Applied nonlinear optics. Courier Dover Publications, 2006.

[90] J. Leuthold, C. Koos, and W. Freude, "Nonlinear silicon photonics," Nat. Photonics, vol. 4, no. 8, pp. 535–544, Jul. 2010.

[91] K. Okamoto, Fundamentals of optical waveguides. Academic press, 2010.

[92] A. W. Snyder, Optical Waveguide Theory. Chapman and Hall, 1983.

[93] J. D. Joannopoulos, S. G. Johnson, J. N. Winn, and R. D. Meade, Photonic crystals: molding the flow of light. Princeton University Press, 2011.

[94] B. Prade, J. Y. Vinet, and A. Mysyrowicz, "Guided optical waves in planar heterostructures with negative dielectric constant," Phys. Rev. B, vol. 44, no. 24, pp. 13556–13572, Dec. 1991.

[95] J. J. Burke, G. I. Stegeman, and T. Tamir, "Surface-polariton-like waves guided by thin, lossy metal films," Phys. Rev. B, vol. 33, no. 8, pp. 5186–5201, Apr. 1986.

[96] J. Dionne, L. Sweatlock, H. Atwater, and a. Polman, "Plasmon slot waveguides: Towards chip-scale propagation with subwavelength-scale localization," Phys. Rev. B, vol. 73, no. 3, p. 035407, Jan. 2006.

[97] N. Feng, M. L. Brongersma, and L. D. Negro, "Metal – dielectric slot-waveguide structures for the propagation of surface plasmon polaritons at 1.55um," IEEE J. Quantum Electron., vol. 43, no. 6, pp. 479–485, 2007.

[98] D. F. P. Pile, D. K. Gramotnev, R. F. Oulton, and X. Zhang, "On long-range plasmonic modes in metallic gaps," Opt. Express, vol. 15, no. 21, pp. 13669–13674, Oct. 2007.

[99] G. Veronis and S. Fan, "Modes of subwavelength plasmonic slot waveguides," J. Light. Technol., vol. 25, no. 9, pp. 2511–2521, 2007.

[100] R. Yang and Z. Lu, "Silicon-on-insulator platform for integration of 3-D nanoplasmonic devices," IEEE Photonics Technol. Lett., vol. 23, no. 22, pp. 1652–1654, Nov. 2011.

[101] D. F. P. Pile and D. K. Gramotnev, "Adiabatic and nonadiabatic nanofocusing of plasmons by tapered gap plasmon waveguides," Appl. Phys. Lett., vol. 89, no. 4, p. 041111, 2006.

[102] C. Delacour, S. Blaize, P. Grosse, J. M. Fedeli, A. Bruyant, R. Salas-Montiel, G. Lerondel, and A. Chelnokov, "Efficient directional coupling between silicon and copper plasmonic nanoslot waveguides: toward metal-oxide-silicon nanophotonics," Nano Lett., vol. 10, no. 8, pp. 2922–2926, Aug. 2010.

[103] Q. Li and M. Qiu, "Structurally-tolerant vertical directional coupling between metal-insulator-metal plasmonic waveguide and silicon dielectric waveguide," Opt. Express, vol. 18, no. 15, pp. 15531–15543, Jul. 2010.

[104] G. Veronis and S. Fan, "Theoretical investigation of compact couplers between dielectric slab waveguides and two-dimensional metal-dielectric-metal plasmonic waveguides," Opt. Express, vol. 15, no. 3, pp. 1211–1221, 2007.

[105] Z. Han, A. Y. Elezzabi, and V. Van, "Experimental realization of subwavelength plasmonic slot waveguides on a silicon platform.," Opt. Lett., vol. 35, no. 4, pp. 502–504, Mar. 2010.

[106] J. Tian, S. Yu, W. Yan, and M. Qiu, "Broadband high-efficiency surface-plasmon-polariton coupler with silicon-metal interface," Appl. Phys. Lett., vol. 95, no. 1, p. 013504, 2009.

[107] L. Chen, J. Shakya, and M. Lipson, "Subwavelength confinement in an integrated metal slot waveguide on silicon," Opt. Lett., vol. 31, no. 14, pp. 2133–2135, Jul. 2006.

[108] L. Y. M. Tobing, L. Tjahjana, and D. Hua Zhang, "Demonstration of low-loss on-chip integrated plasmonic waveguide based on simple fabrication steps on silicon-on-insulator platform," Appl. Phys. Lett., vol. 101, no. 4, p. 041117, 2012.

[109] S. Zhu, G. Q. Lo, and D. L. Kwong, "Nanoplasmonic power splitters based on the horizontal nanoplasmonic slot waveguide," Appl. Phys. Lett., vol. 99, no. 3, p. 031112, 2011.

[110] P. Ginzburg, D. Arbel, and M. Orenstein, "Gap plasmon polariton structure for very efficient microscale-to-nanoscale interfacing," Opt. Lett., vol. 31, no. 22, pp. 3288–3290, 2006.

[111] N.-N. Feng and L. Dal Negro, "Plasmon mode transformation in modulated-index metal-dielectric slot waveguides," Opt. Lett., vol. 32, no. 21, pp. 3086–3088, Dec. 2007.

[112] A. Emboras, R. M. Briggs, A. Najar, S. Nambiar, C. Delacour, P. Grosse, E. Augendre, J. M. Fedeli, B. de Salvo, H. A. Atwater, and R. Espiau de Lamaestre, "Efficient coupler between silicon photonic and metal-insulator-silicon-metal plasmonic waveguides," Appl. Phys. Lett., vol. 101, no. 25, p. 251117, 2012.

[113] X. He, L. Yang, and T. Yang, "Optical nanofocusing by tapering coupled photonic-plasmonic waveguides," Opt. Express, vol. 19, no. 14, pp. 12865–12872, Jul. 2011.

[114] R. Yang, M. A. G. Abushagur, and Z. Lu, "Efficiently squeezing near infrared light into a 21 nm-by-24 nm nanospot," Opt. Express, vol. 16, no. 24, pp. 20142–20148, Dec. 2008.

[115] M. Wu, Z. Han, and V. Van, "Efficient coupler between silicon photonic and metal-insulator-silicon-metal plasmonic waveguides," Opt. Express, vol. 18, no. 11, pp. 11728–36, May 2010.

[116] D. K. Gramotnev and S. I. Bozhevolnyi, "Plasmonics beyond the diffraction limit," Nat. Photonics, vol. 4, no. 2, pp. 83–91, Jan. 2010.

[117] S. Zhu, T. Y. Liow, G. Q. Lo, and D. L. Kwong, "Silicon-based horizontal nanoplasmonic slot waveguides for on-chip integration," Opt. Express, vol. 19, no. 9, pp. 8888–8902, Apr. 2011.

[118] S. Papaioannou, D. Kalavrouziotis, K. Vyrsokinos, J.-C. Weeber, K. Hassan, L. Markey, A. Dereux, A. Kumar, S. I. Bozhevolnyi, M. Baus, T. Tekin, D. Apostolopoulos, H. Avramopoulos, and N. Pleros, "Active plasmonics in WDM traffic switching applications," Sci. Rep., vol. 2, no. 652, pp. 1–9, Jan. 2012.

[119] J. A. Dionne, L. A. Sweatlock, M. T. Sheldon, A. P. Alivisatos, and H. A. Atwater, "Silicon-based plasmonics for on-chip photonics," IEEE J. Sel. Top. Quantum Electron., vol. 16, no. 1, pp. 295–306, 2010.

[120] J. S. Schildkraut, "Long-range surface plasmon electrooptic modulator," Appl. Opt., vol. 27, no. 21, pp. 4587–4590, 1988.

[121] M. Xu, F. Li, T. Wang, J. Wu, L. Lu, L. Zhou, and Y. Su, "Design of an electro-optic modulator based on a silicon-plasmonic hybrid phase shifter," J. Light. Technol., vol. 31, no. 8, pp. 1170–1177, 2013.

[122] A. Melikyan, L. Alloatti, A. Muslija, D. Hillerkuss, P. C. Schindler, J. Li, R. Palmer, D. Korn, S. Muehlbrandt, D. Van Thourhout, B. Chen, R. Dinu, M. Sommer, C. Koos, M. Kohl, W. Freude, and J. Leuthold, "High-speed plasmonic phase modulators," Nat. Photonics, vol. 8, no. 3, pp. 229–233, Feb. 2014.

[123] A. Melikyan, N. Lindenmann, S. Walheim, P. M. Leufke, S. Ulrich, J. Ye, P. Vincze, H. Hahn, T. Schimmel, C. Koos, W. Freude, and J. Leuthold, "Surface plasmon polariton absorption modulator," Opt. Express, vol. 19, no. 9, pp. 8855–8869, Apr. 2011.

[124] J. Leuthold, J. Eckner, E. Gamper, P. A. Besse, and H. Melchior, "Multimode interference couplers for the conversion and combining of

zero- and first-order modes," J. Light. Technol., vol. 16, no. 7, pp. 1228–1239, Jul. 1998.

[125] D. F. G. Gallagher and T. P. Felici, "Eigenmode Expansion Methods for Photonics - Pros and Cons," Proceeding SPIE, vol. 4987, pp. 1–14, 2003.

[126] COMSOL Multiphysics. www.comsol.com.

[127] "RSoft Products." Synopsys' Optical Solutions.

[128] "CST Microwave Studio." .

[129] D. B. Keck and A. R. Tynes, "Spectral response of low-loss optical waveguides," Appl. Opt., vol. 11, no. 7, pp. 1502–1506, Jul. 1972.

[130] "OZ Optics." [Online]. Available: http://www.ozoptics.com/.

[131] C. H. Brown, "Light polarizing body," 2.224.214 12/19401940.

[132] Y. Zhang, H. Lei, and B. Li, "Polarization beam splitter with wide bandwidth in air-hole-based periodic dielectric waveguides," Opt. Commun., vol. 283, no. 10, pp. 2140–2145, May 2010.

[133] W. Zaoui, S. Klinger, W. Vogel, and M. Berroth, "Photonic crystal polarization beam splitter in silicon-on-insulator platform," in 6th Jt. Symp. Opto- Microelectron. Devices Circuit, Berlin, Germany, pp. 75–78, 2010.

[134] X. Ao and S. He, "Polarization beam splitters based on a two-dimensional photonic crystal of negative refraction," Opt. Lett., vol. 30, no. 16, pp. 2152–2154, Aug. 2005.

[135] T. K. Liang and H. K. Tsang, "Integrated polarization beam splitter in high index contrast silicon-on-insulator waveguides," IEEE Photonics Technol. Lett., vol. 17, no. 2, pp. 393–395, 2005.

[136] D. Taillaert, H. Chong, P. I. Borel, L. H. Frandsen, R. M. D. La Rue, and R. Baets, "A compact two-dimensional grating coupler used as a polarization splitter," IEEE Photonics Technol. Lett., vol. 15, no. 9, pp. 1249–1251, 2003.

[137] H. Wu, W. Mo, J. Hou, D. Gao, R. Hao, R. Guo, W. Wu, and Z. Zhou, "Polarizing beam splitter based on a subwavelength asymmetric profile grating," J. Opt., vol. 12, no. 1, p. 015703, Jan. 2010.

[138] C.-Y. Tai, S. H. Chang, and T. Chiu, "Design and analysis of an ultra-compact and ultra-wideband polarization beam splitter based on coupled plasmonic waveguide arrays," IEEE Photonics Technol. Lett., vol. 19, no. 19, pp. 1448–1450, Oct. 2007.

[139] C.-L. Zou, F.-W. Sun, C.-H. Dong, X.-F. Ren, J.-M. Cui, X.-D. Chen, Z.-F. Han, and G.-C. Guo, "Broadband integrated polarization beam splitter with surface plasmon," Opt. Lett., vol. 36, no. 18, pp. 3630–3632, Oct. 2011.

[140] D. J. Thomson, F. Y. Gardes, J.-M. Fedeli, S. Zlatanovic, Y. Hu, B. P. P. Kuo, E. Myslivets, N. Alic, S. Radic, G. Z. Mashanovich, and G. T. Reed, "50Gb/s silicon optical modulator," IEEE Photonics Technol. Lett., vol. 24, no. 4, pp. 234–236, 2012.

[141] J. A. Schuller, E. S. Barnard, W. Cai, Y. C. Jun, J. S. White, and M. L. Brongersma, "Plasmonics for extreme light concentration and manipulation," Nat. Mater., vol. 9, no. 3, pp. 193–204, Mar. 2010.

[142] M. Kauranen and A. V. Zayats, "Nonlinear plasmonics," Nat. Photonics, vol. 6, no. 11, pp. 737–748, Nov. 2012.

[143] Z. Fang, Z. Liu, Y. Wang, P. M. Ajayan, P. Nordlander, and N. J. Halas, "Graphene-antenna sandwich photodetector," Nano Lett., vol. 12, no. 7, pp. 3808–3813, Jul. 2012.

[144] D. Jin, H. Chen, A. Barklund, J. Mallari, G. Yu, E. Miller, and R. Dinu, "EO polymer modulators reliability study," in Int. Soc. Opt. Photonics, p. 75990H–75990H, 2010.

[145] R. Schmogrow, B. Nebendahl, M. Winter, A. Josten, D. Hillerkuss, S. Koenig, J. Meyer, M. Dreschmann, M. Huebner, C. Koos, J. Becker, W. Freude, and J. Leuthold, "Error vector magnitude as a performance

measure for advanced modulation formats," IEEE Photonics Technol. Lett., vol. 24, no. 1, pp. 61–63, Jan. 2012.

[146] X. Zheng, P. Koka, M. O. McCracken, H. Schwetman, J. G. Mitchell, J. Yao, R. Ho, K. Raj, and A. V. Krishnamoorthy, "Energy-efficient error control for tightly coupled systems using silicon photonic interconnects," J. Opt. Commun. Netw., vol. 3, no. 8, pp. A21–A31, Jun. 2011.

[147] D. A. B. Miller, "Energy consumption in optical modulators for interconnects," Opt. Express, vol. 20, no. S2, pp. A293–A308, Mar. 2012.

[148] D. Miller, "Device requirements for optical interconnects to silicon chips," Proc. IEEE, vol. 97, no. 7, pp. 1166–1185, Jul. 2009.

[149] E. Timurdogan, C. M. Sorace-Agaskar, J. Sun, E. Shah Hosseini, A. Biberman, and M. R. Watts, "An ultralow power athermal silicon modulator," Nat. Commun., vol. 5, p. 4008, Jan. 2014.

[150] J. Li, T. P. White, L. O'Faolain, A. Gomez-Iglesias, and T. F. Krauss, "Systematic design of flat band slow light in photonic crystal waveguides," Opt. Express, vol. 16, no. 9, pp. 6227–32, Apr. 2008.

[151] P. Dong, C. Xie, L. Chen, N. K. Fontaine, and Y. Chen, "Experimental demonstration of microring quadrature phase-shift keying modulators.," Opt. Lett., vol. 37, no. 7, pp. 1178–80, Apr. 2012.

[152] S. Zhu, G. Q. Lo, and D. L. Kwong, "Theoretical investigation of silicide Schottky barrier detector integrated in horizontal metal-insulator-silicon-insulator-metal nanoplasmonic slot waveguide," Opt. Express, vol. 19, no. 17, pp. 15843–15854, Aug. 2011.

[153] J. Lee, N. Kaneda, T. Pfau, A. Konczykowska, F. Jorge, J.-Y. Dupuy, and Y.-K. Chen, "Serial 103.125 Gbit/s transmission over 1 km SSMF for low- cost , short-reach optical interconnects," in Opt. Fiber Commun. Conf., San Francisco, CA, USA, no. 1, p. Th5A.5, 2014.

[154] R. Palmer, S. Koeber, D. L. Elder, M. Woessner, W. Heni, D. Korn, M. Lauermann, W. Bogaerts, L. Dalton, W. Freude, J. Leuthold, and C.

Koos, "High-speed, low drive-voltage silicon-organic hybrid modulator based on a binary-chromophore electro-optic material," IEEE J. Light. Technol., vol. 32, no. 16, pp. 2726–2734, 2014.

[155] M. R. Watts, W. A. Zortman, D. C. Trotter, R. W. Young, and A. L. Lentine, "Low-voltage, compact, depletion-mode, silicon Mach–Zehnder modulator," IEEE J. Sel. Top. Quantum Electron., vol. 16, no. 1, pp. 159–164, 2010.

[156] F. Chang, K. Onohara, and T. Mizuochi, "Forward error correction for 100 G transport networks," IEEE Commun. Mag., vol. 48, no. 3, pp. 48–55, 2010.

[157] H. Fukano, T. Yamanaka, M. Tamura, and Y. Kondo, "Very-low-driving-voltage electroabsorption modulators operating at 40 Gbit/s," IEEE J. Light. Technol., vol. 24, no. 5, pp. 2219–2224, 2006.

[158] Y. Chiu, S. Z. Zhang, V. Kaman, J. Piprek, and J. E. Bowers, "High-speed traveling-wave electroabsorption modulators," Proceeding SPIE, vol. 4490, no. 1, pp. 1–10, 2001.

[159] C. Jung, S. Yee, and K. Kuhn, "Electro-optic polymer light modulator based on surface plasmon resonance," Appl. Opt., vol. 34, no. 6, pp. 946–949, Mar. 1995.

[160] C. Jung, S. Yee, and K. Kuhn, "Integrated optics waveguide modulator based on surface plasmon resonance," IEEE J. Light. Technol., vol. 12, no. 10, pp. 1802–1806, 1994.

[161] W. Kuo and M. Chen, "Simulation study of surface-plasmon-polariton electro-optic light modulator based on a polymer grating coupler," Opt. Lett., vol. 34, no. 24, pp. 3812–3814, 2009.

[162] Z. Wu, R. L. Nelson, J. W. Haus, and Q. Zhan, "Plasmonic electro-optic modulator design using a resonant metal grating.," Opt. Lett., vol. 33, no. 6, pp. 551–3, Mar. 2008.

[163] O. Solgaard, F. Ho, J. I. Thackara, and D. M. Bloom, "High frequency attenuated total internal reflection light modulator," Appl. Phys. Lett., vol. 61, no. 21, pp. 2500–2502, 1992.

[164] M. J. Dicken, L. A. Sweatlock, D. Pacifici, H. J. Lezec, K. Bhattacharya, and H. A. Atwater, "Electrooptic modulation in thin film barium titanate plasmonic interferometers," Nano Lett., vol. 8, no. 11, pp. 4048–4052, 2008.

[165] A. Melikyan, T. Vallaitis, N. Lindenmann, T. Schimmel, W. Freude, and J. Leuthold, "A surface plasmon polariton absorption modulator," in Conf. Lasers Electro-Optics, Baltimore, MD, USA, p. JThE77, 2010.

[166] M. L. Nesterov, A. V Kats, and S. K. Turitsyn, "Extremely short-length surface plasmon resonance devices," Opt. Express, vol. 16, no. 25, pp. 20227–20240, Dec. 2008.

[167] F. Neumann, Y. Genenko, C. Melzer, S. Yampolskii, and H. von Seggern, "Self-consistent analytical solution of a problem of charge-carrier injection at a conductor/insulator interface," Phys. Rev. B, vol. 75, no. 20, p. 205322, May 2007.

[168] F. Michelotti, L. Dominici, E. Descrovi, N. Danz, and F. Menchini, "Thickness dependence of surface plasmon polariton dispersion in transparent conducting oxide films at 1.55um," Opt. Lett., vol. 34, no. 6, pp. 839–841, Mar. 2009.

[169] S. Dasgupta, M. Lukas, K. Dössel, R. Kruk, and H. Hahn, "Electron mobility variations in surface-charged indium tin oxide thin films," Phys. Rev. B, vol. 80, no. 8, p. 085425, Aug. 2009.

[170] A. V. Krasavin and A. V. Zayats, "Photonic signal processing on electronic scales: electro-optical field-effect nanoplasmonic modulator," Phys. Rev. Lett., vol. 109, no. 5, p. 053901, Jul. 2012.

[171] J. Peatross, S. A. Glasgow, and M. Ware, "Average energy flow of optical pulses in dispersive media," Phys. Rev. Lett., vol. 84, no. 11, pp. 2370–2373, Mar. 2000.

[172] J.-M. Jin, The finite element method in electromagnetics. Wiley New York, 2002.

[173] T. J. Davis, "Surface plasmon modes in multi-layer thin-films," Opt. Commun., vol. 282, no. 1, pp. 135–140, Jan. 2009.

[174] C. Rhodes, M. Cerruti, a. Efremenko, M. Losego, D. E. Aspnes, J.-P. Maria, and S. Franzen, "Dependence of plasmon polaritons on the thickness of indium tin oxide thin films," J. Appl. Phys., vol. 103, no. 9, p. 093108, 2008.

[175] H. S. Kwok, X. W. Sun, and D. H. Kim, "Pulsed laser deposited crystalline ultrathin indium tin oxide films and their conduction mechanisms," Thin Solid Films, vol. 335, no. 1, pp. 299–302, 1998.

[176] J. Dionne, L. Sweatlock, H. Atwater, and A. Polman, "Planar metal plasmon waveguides: frequency-dependent dispersion, propagation, localization, and loss beyond the free electron model," Phys. Rev. B, vol. 72, no. 7, p. 075405, 2005.

[177] "Www.gigoptix.com." 2010.

Acknowledgements

This work would never have been possible without the help and the support of all the great people that I had the chance to work with. I would like to express my deepest appreciation to all that have shared their time, knowledge and experience with me.

Foremost, my sincerest gratitude goes to my advisors Prof. Juerg Leuthold and Prof. Wolfgang Freude for all the time and resources that they have invested in my work. I owe my deepest gratitude to Prof. J. Leuthold for all the motivation and the inspiration that he has given me during my PhD studies. I am grateful to him for all the discussions and brainstorming sessions that helped me to find new and better ways towards the big demonstration. My deepest appreciation goes also to Prof. Wolfgang Freude for all his invaluable teachings and constructive criticism which has always motivated me to learn new things and try new things. I am grateful to Prof. W. Freude for all the effort that he put in improving my manuscripts and my PhD thesis.

Special thanks go to PD Dr. Manfred Kohl for supporting my work at the Institute of Microstructure Technology (IMT) and giving me an access to the dry etching equipment. Martin Sommer I thank for giving me permission to use his oxidation oven and for support with the NAVOLCHI project management.

I would like to express my special thanks to Prof. Christian Koos for initiating the fabrication run for the passive silicon platform of the plasmonic phase modulators. I am also grateful to him for his insightful comments and suggestions on my manuscripts.

I am grateful to all my colleagues at the Institute of Photonics and Quantum Electronics (IPQ) for their contributions to my work and for the pleasant environment that they have created. In particular, I thank Luca Alloatti for helping me with the poling of the electro-optic material. David Hillerkuss I thank for designing the experimental setup for performing data modulation experiments. Dietmar Korn I thank for his support in designing silicon photonic chips. Philipp Schindler I thank for helping me in all my system experiments. Robert Palmer I thank for all the creative and insightful discussions. Jingshi Li I thank for providing me the best delay interferometer in the world. Nicole Lindenmann I thank for her vital effort and support in the realization of the plasmonic absorption modulators. Rene Schmogrow I thank for all the discussions that we had on digital communications. I thank my former colleagues Rene Bonk, Swen König, Röger Moritz for assisting me in the characterization laboratory at IPQ. Sascha Mühlbrandt I thank for his help with the NAVOLCHI project management at IMT.

I am thankful to Prof. Uli Lemmer for granting access to the cleanroom of Light Technology Institute (LTI). I thank Thorsten Feldmann and Christian Kayser of LTI for their help in the cleanroom during the first period of my PhD studies.

My sincere thanks also go to Prof. Gerd Schön for his important feedback on the Thomas-Fermi screening.

I am grateful to the entire staff of the Institute of Microstructure Technology. In particular, I thank Alban Muslija, Dr. Daniel Häringer, Peter-Jürgen Jakobs, Andreas Bacher, Dr. Markus Guttmann, and Dr. Holger Moritz for being so supportive, flexible and patient.

I am grateful to Prof. Thomas Schimmel, Prof. Horst Hahn and Dr. Stefan Walheim of the Institute of Nanotechnology for their collaboration in fabricating the plasmonic absorption modulators. I thank Paul Vincze for the high-quality metal layers that he deposited for me. Philipp Leufke I thank for sputtering indium tin oxide layers on our plasmonic samples. I would also like to express my gratitude to Christian Benz and Simone Dehm for the insightful introductions to the Raith electron beam lithography system.

I thank Dr. Sven Ulrich and Dr. Jian Ye of the Institute of Material Science I (KIT) for the high quality insulator layers that they deposited on our samples.

I thank Prof. Gernot Goll for granting access to the entire manufacturing facility at the Center for Functional Nanostructures (CFN). I am grateful for all the assistance given by the CFN staff. In particular, I thank Stefan Kühn, Silvia Diewald, Patrice Brenner, and Simon Hettler for their support and their willingness to try out new things.

I thank Dr. Raluca Dinu and Dr. Baoquan Chen of GigOptix for sharing their electro-optic material with us.

I thank Prof. Dries Van Thourhout of Ghent University – imec for providing an easy access to the ePIXfab technology.

I would like to thank all my students, Kira Köhnle, Dominik Meier, Claus Gärtner, Christian Haffner, Levon Yeghiazaryan, Sandeep Ummethala as well as Radwanul Hasan Siddique for their collaboration and for their enthusiasm for plasmonics.

I thank Hans Bürger and Manfred Hirsch for building the mechanical parts of the plasmonic setup. The accuracy and the speed of their work have always fascinated me. Oswald Speck, Florian Rupp and Martin Winkeler I thank for their technical support. I thank our secretaries Bernadette Lehmann, Andrea Riemensperger and Angelika Olbrich for their important work and help in administrative matters.

My special thanks go to Seán Ó Dúill for carefully going through this thesis and helping to improve grammar and wording.

My special thanks go to Prof. Vanik Mkrtchian for believing in me.

I am thankful to my parents Zareh and Venera as well as my sister Urartuhi and brother Menua for their unconditional support. Finally, I thank my wife Lilit for her understanding, love and support.

This work was supported by the EU research projects NAVOLCHI and SOFI, the Karlsruhe School of Optics & Photonics (KSOP), the Center for Functional Nanostructures (CFN) and the Karlsruhe Nano Micro Facility (KNMF). Silicon-on-insulator waveguides were fabricated by imec in the framework of the ePIXfab.

List of Own Publications

Journal Papers

[J1] **A. Melikyan**, M. Kohl, M. Sommer, C. Koos, W. Freude, and J. Leuthold, "Photonic-to-plasmonic mode converter," *Opt. Lett.,* vol. 39, no. 12, pp. 3488-3491, June 2014

[J2] **A. Melikyan**, L. Alloatti, A. Muslija, D. Hillerkuss, P. C. Schindler, J. Li, R. Palmer, D. Korn, S. Muehlbrandt, D. Van Thourhout, B. Chen, R. Dinu, M. Sommer, C. Koos, M. Kohl, W. Freude, and J. Leuthold, "High-speed plasmonic phase modulators," ***Nat. Photonics***, vol. 8, no. 2, pp. 229–233, Feb. 2014.

[J3] **A. Melikyan**, K. Koehnle, M. Lauermann, R. Palmer, S. Koeber, S. Muehlbrandt, P. C. Schindler, D. L. Elder, S. Wolf, W. Heni, C. Haffner, Y. Fedoryshyn, D. Hillerkuss, M. Sommer, L. R. Dalton, D. Van Thourhout, W. Freude, M. Kohl, J. Leuthold, and C. Koos, "Plasmonic-organic hybrid (POH) modulators for OOK and BPSK signaling at 40 Gbit/s," *Opt. Express*, vol. 23, no. 8, pp. 9938–9946, Apr. 2015.

[J4] **A. Melikyan**, N. Lindenmann, S. Walheim, P. M. Leufke, S. Ulrich, J. Ye, P. Vincze, H. Hahn, T. Schimmel, C. Koos, W. Freude, and J. Leuthold, "Surface plasmon polariton absorption modulator.," *Opt. Express*, vol. 19, no. 9, pp. 8855–8869, Apr. 2011.

[J5] J. Leuthold, C. Hoessbacher, S. Muehlbrandt, **A. Melikyan**, M. Kohl, C. Koos, W. Freude, V. Dolores-Calzadilla, M. Smit, I: Suarez, J. Martinez-Pastor, E. P. Fitrakis, I. Tomkos, "Plasmonic Communications: Light on a Wire," *Opt. and Photonics News*, vol. 24, pp. 28–35, 2013.

[J6] D. Hillerkuss, R. Schmogrow, M. Meyer, S. Wolf, M. Jordan, P. Kleinow, N. Lindenmann, P. C. Schindler, **A. Melikyan**, X. Yang, S. Ben-Ezra, B. Nebendahl, M. Dreschmann, J. Meyer, F. Parmigiani, P. Petropoulos, B. Resan, A. Oehler, K. Weingarten, L. Altenhain, T. Ellermeyer, M. Moeller,

M. Huebner, J. Becker, C. Koos, W. Freude, J. Leuthold, "Single-laser 32.5 Tbit / s Nyquist WDM transmission.," *J. Opt. Comm. Netw.*, vol. 4, no. 4, pp. 715–723, March 2012.

Conference Contributions

[C1] **A. Melikyan**, C. Gaertner, K. Köhnle, A. Muslija, M. Sommer, M. Kohl, C. Koos, W. Freude, and J. Leuthold, "Integrated wire grid polarizer and plasmonic polarization beam splitter," in *Proc. of Optical Fiber Commun. Conf.*, Los Angeles, CA, USA, p. OW1E.3, 2012.

[C2] **A. Melikyan**, L. Alloatti, A. Muslija, D. Hillerkuss, P. Schindler, J. Li, R. Palmer, D. Korn, S. Muehlbrandt, D. Van Thourhout, B. Chen, R. Dinu, M. Sommer, C. Koos, M. Kohl, W. Freude, and J. Leuthold, "Surface plasmon polariton high-speed modulator," in *Conf. Lasers Electro-Optics*, San Jose, CA, USA, p. CTh5D.2, 2013. **(Postdeadline Paper)**

[C3] **A. Melikyan**, T. Vallaitis, N. Lindenmann, T. Schimmel, W. Freude, and J. Leuthold, "A surface plasmon polariton absorption modulator.," in *Conf. Lasers Electro-Optics*, Baltimore, MD, USA, p. JThE77, 2010.

[C4] **A. Melikyan**, M. Sommer, A. Muslija, M. Kohl, S. Muehlbrandt, A. Mishra, V. Calzadilla, Y. Justo, J. P. Martinez-Pastor, I. Tomkos, A. Scandurra, D. Van Thourhout, Z. Hens, M. Smit, W. Freude, C. Koos, and J. Leuthold, "Chip-to-chip plasmonic interconnects and the activities of EU project NAVOLCHI," in *Transparent Opt. Networks (ICTON), 2012 14th Int. Conf.*, pp. 1–3, 2012 **(invited).**

[C5] W. Freude, L. Alloatti, **A. Melikyan**, R. Palmer, D. Korn, N. Lindenmann, T. Vallaitis, D. Hillerkuss, J. Li, A. Barklund, R. Dinu, J. Wieland, M. Fournier, J.-M. Fedeli, S. Walheim, P. Leufke, S. Ulrich, J. Ye, P. Vincze, H. Hahn, H. Yu, W. Bogaerts, P. Dumon, R. Baets, B. Breiten, F. Diederich, M. T. Beels, I. Biaggio, T. Schimmel, C. Koos, and J. Leuthold, "Nonlinear Optics on the Silicon Platform," in *Proc. of Optical Fiber Commun. Conf.*, p. OTh3H.6, 2012 **(invited)**.

[C6] W. Freude, L. Alloatti, D. Korn, M. Lauermann, **A. Melikyan**, R. Palmer, J. Pfeifle, P. C. Schindler, C. Weimann, R. Dinu, J. Bolten, T. Wahlbrink, M. Waldow, S. Walheim, P. Leufke, S. Ulrich, J. Ye, P. Vincze, H. Hahn, H. Yu, W. Bogaerts, V. Brasch, T. Herr, R. Holzwarth, K. Hartinger, C. Stamatiadis, M. O'Keefe, L. Stampoulidis, L. Zimmermann, R. Baets, T. Schimmel, I. Tomkos, K. Petermann, T. J. Kippenberg, C. G. Koos, and J. Leuthold, "Nonlinear Nano-Photonics," in *Front. Opt. 2013*, p. FTu4C.6, 2013 **(invited).**